MW00333717

Science Starters: Elementary Physical & Earth Science

Master Books®
Curriculum

First printing: April 2013
Third printing: August 2018

Master Books®, P.O. Box 726, Green Forest, AR 72638

Master Books® is a division of the New Leaf Publishing Group, Inc.

ISBN: 978-0-89051-969-1
ISBN: 978-1-61458-580-0 (digital)

Unless otherwise noted, Scripture quotations are from the New King James Version of the Bible.

Printed in the United States of America

Please visit our website for other great titles:
www.masterbooks.com

For information regarding author interviews,
please contact the publicity department at (870) 438-5288.

"I'm loving this whole line so much. It's changed our homeschool for the better!

—Amy ★★★★★

"Your reputation as a publisher is stellar. It is a blessing knowing anything I purchase from you is going to be worth every penny!

—Cheri ★★★★★

"Last year we found Master Books and it has made a HUGE difference.

—Melanie ★★★★★

"We love Master Books and the way it's set up for easy planning!

—Melissa ★★★★★

"You have done a great job. MASTER BOOKS ROCKS!

—Stephanie ★★★★★

"Physically high-quality, Biblically faithful, and well-written.

—Danika ★★★★★

"Best books ever. Their illustrations are captivating and content amazing!

—Kathy ★★★★★

Affordable
Flexible
Faith Building

Table of Contents

Using This Teacher Guide

Features: The suggested weekly schedule enclosed has easy-to-manage lessons that guide the reading, worksheets, and all assessments. The pages of this guide are perforated and three-hole punched so materials are easy to tear out, hand out, grade, and store. Teachers are encouraged to adjust the schedule and materials needed in order to best work within their unique educational program.

Lesson Scheduling: Students are instructed to read the pages in their book and then complete the corresponding section provided by the teacher. Assessments that may include worksheets, activities, quizzes, and tests are given at regular intervals with space to record each grade. Space is provided on the weekly schedule for assignment dates, and flexibility in scheduling is encouraged. Teachers may adapt the scheduled days per each unique student situation. As the student completes each assignment, this can be marked with an "X" in the box.

🕐	**Approximately 30 to 45 minutes per lesson, two to three days a week**
🔑	**Includes answer keys for quizzes and tests.**
📄	**Quizzes and tests are included to help reinforce learning and provide assessment opportunities.**
🔁	**Designed for grades 4 to 6 in a one-year course**

Course Objectives: Students completing this course will

- Learn how to determine the speed and motion of favorite toys
- Create a catapult and experience the mechanics of pulleys
- Examine natural occurrences such as mountains, volcanoes, rocks, minerals, crystals, water, and dirt
- Discover why friction creates heat

- Use household items such as hard boiled eggs, oranges, measuring cups, maps, clay and markers to see these scientific truths will come to life.
- Learn how to determine the speed and motion of favorite toys, create a catapult and experience the mechanics of pulleys, set up a floating pencil race, and discover why friction creates heat.

Course Description

The *Investigate the Possibilities* curriculum has been developed with the following learning progression:

Engage - Students make a note of what they know or have experienced about the topic.

Investigate - Students will follow the instructions and make observations of what happens.

Explain - Students will begin to understand the science behind what they observed in the investigation.

Apply - Here, the understanding of the investigation is related to other situations and ideas.

Expand -Each investigation also includes a few "Dig Deeper" projects to further understanding.

Assess- Students explain what they have learned.

Elementary physical science and earth science comes alive through this activities-driven science course that ignites a sense of curiosity about the wonderful world God has made. Concepts are introduced in an engaging way by highlighting the science behind kids at play, like roller-skating, skateboarding, and even running. By guiding students through these easy-to-understand investigations, they learn to observe and relate what they have personally observed in detail. The learning progression helps students engage, investigate, explain, apply, expand, and assess the scientific principles, and is filled with helpful images, diagrams, and inexpensive activities. Students discover why caves and sinkholes form, what is in the soil we walk on every day, how warning signs are present prior to volcanic eruptions, what tests can be used to identify rocks, and more. This comprehensive series makes the study of God's creation both enjoyable and educational!

Calculating a Final Grade

Calculate the Average of the student's Activities & Observations grades.

Divide the average by 3 _____

Calculate the Average of the student's Questions & Quizzes grades.

Divide the average by 3 _____

Calculate the Average of the student's Projects, Contest & Dig Deeper grades.

Divide the average by 3 _____

Add up the numbers for the Final Grade: _____

Suggested Optional Science Lab

There are a variety of companies that offer science labs that complement our courses. These items are only suggestions, not requirements, and they are not included in the daily schedule. We have tried to find materials that are free of evolutionary teaching, but please review any materials prior to presentation. The following items are available from www.HomeTrainingTools.com.

RM-GEOBAG Geology Field Trip in a Bag

RM-ROCKMIN Rocks & Minerals of the U.S. Basic Set

First Semester Suggested Daily Schedule

Date	Day	Assignment	Due Date	✓	Grade
		First Semester-First Quarter — *Forces & Motion*			
Week 1	Day 1				
	Day 2	Investigation #1: Wind-up Walking Toys Read Pages 4-7 • *Forces & Motion* (FM) Complete Page S4 • Student Journal (SJ)			
	Day 3				
	Day 4	Investigation #1: Wind-up Walking Toys Read Pages 8-9 • (FM) • Complete Page S5 • (SJ)			
	Day 5				
Week 2	Day 6				
	Day 7	Investigation #2: Which Way Did It Go? Read Pages 10-11 • (FM) • Complete Page S6 • (SJ)			
	Day 8				
	Day 9	Investigation #2: Which Way Did It Go? Read Pages 12-13 • (FM) • Complete Page S7 • (SJ)			
	Day 10				
Week 3	Day 11				
	Day 12	Investigation #3: Investigating Friction Read Pages 14-15 • (FM) • Complete Page S8 • (SJ)			
	Day 13				
	Day 14	Investigation #3: Investigating Friction Read Pages 16-17 • (FM) • Complete Page S9 • (SJ)			
	Day 15				
Week 4	Day 16				
	Day 17	Investigation #4: Friction — Does It Rub You... Read Pages 18-19 • (FM) • Complete Page S10 • (SJ)			
	Day 18				
	Day 19	Investigation #4: Friction — Does It Rub You... Read Pages 20-21 • (FM) • Complete Page S11 • (SJ)			
	Day 20				
Week 5	Day 21	**Forces & Motion Investigations 1-4 Quiz 1** **Level 1** Page 17 • **Level 2** Page 29 • Lesson Plan (LP)			
	Day 22				
	Day 23	Investigation #5: That's Heavy, Dude Read Pages 22-23 • (FM) • Complete Page S12 • (SJ)			
	Day 24				
	Day 25	Investigation #5: That's Heavy, Dude Read Pages 24-25 • (FM) • Complete Page S13 • (SJ)			

Date	Day	Assignment	Due Date	✓	Grade
	Day 26	Investigation #6: Floating Pencil Race Read Pages 26-27 • (FM) • Complete Page S14 • (SJ)			
	Day 27				
Week 6	Day 28	Investigation #6: Floating Pencil Race Read Pages 28-29 • (FM) • Complete Page S15 • (SJ)			
	Day 29				
	Day 30	Investigation #7: What Floats Your Boat? Read Pages 30-31 • (FM) • Complete Page S16 • (SJ)			
	Day 31	Investigation #7: What Floats Your Boat? Read Pages 32-33 • (FM) • Complete Page S17 • (SJ)			
	Day 32				
Week 7	Day 33	Investigation #8: Giving Airplanes a Lift Read Pages 34-35 • (FM) • Complete Page S18 • (SJ)			
	Day 34				
	Day 35	Investigation #8: Giving Airplanes a Lift Read Pages 36-37 • (FM) • Complete Page S19 • (SJ)			
	Day 36				
	Day 37	**Forces & Motion Investigations 5-8 Quiz 2** **Level 1** Page 19 • **Level 2** Page 31 • (LP)			
Week 8	Day 38				
	Day 39	Investigation #9: Crash Test Dummes Read Pages 38-39 • (FM) • Complete Page S20 • (SJ)			
	Day 40				
	Day 41	Investigation #9: Crash Test Dummes Read Pages 40-41 • (FM) • Complete Page S21 • (SJ)			
	Day 42				
Week 9	Day 43	Investigation #10: Cars and Ramps Read Pages 42-43 • (FM) • Complete Page S22 • (SJ)			
	Day 44				
	Day 45	Investigation #10: Cars and Ramps Read Pages 44-45 • (FM) • Complete Page S23 • (SJ)			
First Semester-Second Quarter — *Forces & Motion*					
	Day 46	Investigation #11: The Mighty Conquering Catapults Read Pages 46-47 • (FM) • Complete Page S24 • (SJ)			
	Day 47				
Week 1	Day 48	Investigation #11: The Mighty Conquering Catapults Read Pages 48-49 • (FM) • Complete Page S25 • (SJ)			
	Day 49				
	Day 50	Investigation #11: The Mighty Conquering Catapults — Isaac Newton • Read Pages 50-51 • (FM)			

Date	Day	Assignment	Due Date	✓	Grade
Week 2	Day 51	Investigation #12: Round and Round without Stopping Read Pages 52-53 • (FM) • Complete Page S26 • (SJ)			
	Day 52				
	Day 53	Investigation #12: Round and Round without Stopping Read Pages 54-55 • (FM) • Complete Page S27 • (SJ)			
	Day 54				
	Day 55	Investigation #13: Roller Derby with Flour Read Pages 56-57 • (FM) • Complete Page S28 • (SJ)			
Week 3	Day 56	Investigation #13: Roller Derby with Flour Read Pages 58-59 • (FM) • Complete Page S29 • (SJ)			
	Day 57				
	Day 58	Investigation #14: Balloon Jet Propulsion Read Pages 60-61 • (FM) • Complete Page S30 • (SJ)			
	Day 59				
	Day 60	Investigation #14: Balloon Jet Propulsion Read Pages 62-63 • (FM) • Complete Page S31 • (SJ)			
Week 4	Day 61	**Forces & Motion Investigations 9-14 Quiz 3** **Level 1** Page 21 • **Level 2** Page 33 • (LP)			
	Day 62				
	Day 63	Investigation #15: Balancing Act with a Stick Read Pages 64-65 • (FM) • Complete Page S32 • (SJ)			
	Day 64				
	Day 65	Investigation #15: Balancing Act with a Stick Read Pages 66-67 • (FM) • Complete Page S33 • (SJ)			
Week 5	Day 66				
	Day 67	Investigation #16: Spinning Tops Read Pages 68-69 • (FM) • Complete Page S34 • (SJ)			
	Day 68				
	Day 69	Investigation #16: Spinning Tops Read Pages 70-71 • (FM) • Complete Page S35 • (SJ)			
	Day 70				
Week 6	Day 71				
	Day 72	Investigation #17: He Ain't Heavy, He's Just My... Read Pages 72-73 • (FM) • Complete Page S36 • (SJ)			
	Day 73				
	Day 74	Investigation #17: He Ain't Heavy, He's Just My... Read Pages 74-75 • (FM) • Complete Page S37 • (SJ)			
	Day 75				
Week 7	Day 76				
	Day 77	Investigation #18: How Do Like Your Pulleys? Read Pages 76-77 • (FM) • Complete Page S38 • (SJ)			
	Day 78				
	Day 79	Investigation #18: How Do Like Your Pulleys? Read Pages 78-79 • (FM) • Complete Page S39 • (SJ)			
	Day 80				

Date	Day	Assignment	Due Date	✓	Grade
Week 8	Day 81	Investigation #19: And the Wheel Goes Round Read Pages 80-81 • (FM) • Complete Page S40 • (SJ)			
	Day 82				
	Day 83	Investigation #19: And the Wheel Goes Round Read Pages 82-83 • (FM) • Complete Page S41 • (SJ)			
	Day 84				
	Day 85	Investigation #20: If It Doesn't Move, How Can It Be... Read Pages 84-85 • (FM) • Complete Page S42 • (SJ)			
Week 9	Day 86	Investigation #20: If It Doesn't Move, How Can It Be... Read Pages 86-87 • (FM) • Complete Page S43 • (SJ)			
	Day 87				
	Day 88	**Forces & Motion Investigations 15-20 Quiz 4** **Level 1** Page 23 • **Level 2** Page 37 • (LP)			
	Day 89				
	Day 90	**Forces & Motion Investigations 1-20 Test 1** **Level 1** Page 25 • **Level 2** Page 39 • (LP)			
		Mid-Term Grade			

Second Semester Suggested Daily Schedule

Date	Day	Assignment	Due Date	✓	Grade
		Second Semester-Third Quarter — *The Earth*			
	Day 91				
	Day 92	Investigation #1: Orange You Going to Map the Earth? Read Pages 4-7 • *The Earth* (TE) Complete Page S1 • Student Journal (SJ)			
Week 1	Day 93				
	Day 94	Investigation #1: Orange You Going to Map the Earth? Read Pages 8-9 • (TE) • Complete Page S2 • (SJ)			
	Day 95				
	Day 96				
	Day 97	Investigation #2: Composition of the Earth Read Pages 10-11 • (TE) • Complete Page S3 • (SJ)			
Week 2	Day 98				
	Day 99	Investigation #2: Composition of the Earth Read Pages 12-13 • (TE) • Complete Page S4 • (SJ)			
	Day 100				
	Day 101				
	Day 102	Investigation #3: Why Is Everything Moving? Read Pages 14-15 • (TE) • Complete Page S5 • (SJ)			
Week 3	Day 103				
	Day 104	Investigation #3: Why Is Everything Moving? Read Pages 16-17 • (TE) • Complete Page S6 • (SJ)			
	Day 105				
	Day 106	Investigation #4: Earthquake Read Pages 18-19 • (TE) • Complete Page S7 • (SJ)			
	Day 107				
Week 4	Day 108	Investigation #4: Earthquake Read Pages 20-21 • (TE) • Complete Page S8 • (SJ)			
	Day 109				
	Day 110	**The Earth Investigations 1-4 Quiz 1** **Level 1** Page 45 • **Level 2** Page 59 • (LP)			
	Day 111				
	Day 112	Investigation #5: Living with Earthquakes Read Pages 22-23 • (TE) • Complete Page S9 • (SJ)			
Week 5	Day 113				
	Day 114	Investigation #5: Living with Earthquakes Read Pages 24-25 • (TE) • Complete Page S10 • (SJ)			
	Day 115				

Date	Day	Assignment	Due Date	✓	Grade
	Day 116				
	Day 117	Investigation #6: Volcanoes Read Pages 26-27 • (TE) • Complete Page S11 • (SJ)			
Week 6	Day 118				
	Day 119	Investigation #6: Volcanoes Read Pages 28-31 • (TE) • Complete Page S12 • (SJ)			
	Day 120				
	Day 121	Investigation #7: Mountains (Folding and Faulting) Read Pages 32-33 • (TE) • Complete Page S13 • (SJ)			
	Day 122				
Week 7	Day 123	Investigation #7: Mountains (Folding and Faulting) Read Pages 34-35 • (TE) • Complete Page S14 • (SJ)			
	Day 124				
	Day 125	Investigation #8: Pardon the Intrusion Read Pages 36-37 • (TE) • Complete Page S15 • (SJ)			
	Day 126	Investigation #8: Pardon the Intrusion Read Pages 38-39 • (TE) • Complete Page S16 • (SJ)			
	Day 127				
Week 8	Day 128	**The Earth Investigations 5-8 Quiz 2** **Level 1** Page 47 • **Level 2** Page 61 • (LP)			
	Day 129				
	Day 130	Investigation #9: Mapping a Mountain Read Pages 40-41 • (TE) • Complete Page S17 • (SJ)			
	Day 131	Investigation #9: Mapping a Mountain Read Pages 42-43 • (TE) • Complete Page S18 • (SJ)			
	Day 132				
Week 9	Day 133	Investigation #10: Growing Crystals Read Pages 44-45 • (TE) • Complete Page S19 • (SJ)			
	Day 134				
	Day 135	Investigation #10: Growing Crystals Read Pages 46-47 • (TE) • Complete Page S20 • (SJ)			
Second Semester-Fourth Quarter — *The Earth*					
	Day 136				
	Day 137	Investigation #11: Minerals Read Pages 48-49 • (TE) • Complete Page S21 • (SJ)			
Week 1	Day 138				
	Day 139	Investigation #11: Minerals Read Pages 50-51 • (TE) • Complete Page S22 • (SJ)			
	Day 140				

Date	Day	Assignment	Due Date	✓	Grade
Week 2	Day 141	Investigation #12: Rocks That Fizz Read Pages 52-53 • (TE) • Complete Page S23 • (SJ)			
	Day 142				
	Day 143	Investigation #12: Rocks That Fizz Read Pages 54-55 • (TE) • Complete Page S24 • (SJ)			
	Day 144				
	Day 145	**The Earth Investigations 9-12 Quiz 3** **Level 1** Page 49 • **Level 2** Page 63 • (LP)			
Week 3	Day 146				
	Day 147	Investigation #13: Rocks Have an ID Read Pages 56-57 • (TE) • Complete Page S25 • (SJ)			
	Day 148				
	Day 149	Investigation #13: Rocks Have an ID Read Pages 58-59 • (TE) • Complete Page S26 • (SJ)			
	Day 150				
Week 4	Day 151				
	Day 152	Investigation #14: How Little, Tiny Things Settle... Read Pages 60-61 • (TE) • Complete Page S27 • (SJ)			
	Day 153				
	Day 154	Investigation #14: How Little, Tiny Things Settle... Read Pages 62-63 • (TE) • Complete Page S28 • (SJ)			
	Day 155				
Week 5	Day 156	Investigation #15: How Rocks and Dirt Catch a Ride Read Pages 64-65 • (TE) • Complete Page S29 • (SJ)			
	Day 157				
	Day 158	Investigation #15: How Rocks and Dirt Catch a Ride Read Pages 66-67 • (TE) • Complete Page S30 • (SJ)			
	Day 159				
	Day 160	Investigation #16: Physical and Chemical Weathering Read Pages 68-69 • (TE) • Complete Page S31 • (SJ)			
Week 6	Day 161	Investigation #16: Physical and Chemical Weathering Read Pages 70-71 • (TE) • Complete Page S32 • (SJ)			
	Day 162				
	Day 163	**The Earth Investigations 13-16 Quiz 4** **Level 1** Page 51 • **Level 2** Page 65 • (LP)			
	Day 164				
	Day 165	Investigation #17: Holes in Rocks Read Pages 72-73 • (TE) • Complete Page S33 • (SJ)			
Week 7	Day 166	Investigation #17: Holes in Rocks Read Pages 74-75 • (TE) • Complete Page S34 • (SJ)			
	Day 167				
	Day 168	Investigation #18: Holes in Rocks Read Pages 76-77 • (TE) • Complete Page S35 • (SJ)			
	Day 169				
	Day 170	Investigation #18: Holes in Rocks Read Pages 78-79 • (TE) • Complete Page S36 • (SJ)			

Date	Day	Assignment	Due Date	✓	Grade
Week 8	Day 171	Investigation #19: Glaciers Read Pages 80-81 • (TE) • Complete Page S37 • (SJ)			
	Day 172				
	Day 173	Investigation #19: Glaciers Read Pages 82-83 • (TE) • Complete Page S38 • (SJ)			
	Day 174				
	Day 175	Investigation #20: Toiling in the Soil Read Pages 84-85 • (TE) • Complete Page S39 • (SJ)			
Week 9	Day 176	Investigation #20: Toiling in the Soil Read Pages 86-87 • (TE) • Complete Page S40 • (SJ)			
	Day 177				
	Day 178	**The Earth Investigations 17-20 Quiz 5** **Level 1** Page 53 • **Level 2** Page 67 • (LP)			
	Day 179				
	Day 180	**The Earth Investigations 1-20 Test 1** **Level 1** Page 55 • **Level 2** Page 69 • (LP)			
		Final Grade			

Quizzes and Test

for Use with

Forces & Motion

Testing:

This series is appropriate for elementary students from 3rd to 6th grades. Because of this, we have included quizzes and tests in two different levels, which you can choose from based on your child's abilities and understanding of the concepts in the course.

Level 1: suggested for younger ages or those who struggle with application of the concepts beyond just definitions and basic concepts

Level 2: suggested for older ages or those who can both grasp the scientific concepts and apply them consistently

Permission to Copy

Choose answers from these terms.
All the terms may not be used and some may be used more than once:

centimeters	engineer	frame of reference	friction	heat
increase	inertia	no	north	reduce
rolling friction	scientist	second	sliding friction	south
streamlined	time	yes		

Fill in the Blank: Each question is worth 5 points.

1. In order to find something's speed, you need to divide the distance by the _____.

2. Does this graph of a runner's speed show that the runner kept on going faster and faster? _____

3. Suppose a bug's speed is 2 cm/s. This means it can travel 2 _____ per _____.

4. In order to tell how something is moving, you need a _____.

5. A flat boat is moving north at 10 miles per hour. A walking toy on the boat is moving south at 5 miles per hour. Would someone watching the walking toy from the bank of the river see it moving north or south?

6. Is Darwin's idea that all living things evolved from a one-celled organism represented by a forest of trees?

7. What force causes your tennis shoes to wear out?

8. What force helps you to walk without falling down?

9. Friction always produces what form of energy?

10. Ball bearings and roller bearings are used to _____ friction.

11. Is it possible for the same set of facts to be interpreted in more than one way?

12. Many birds have a special shape that lets them fly smoothly through the air with little friction. What is this shape called?

13. Does a force have to be applied to an object to cause it to start moving?

14. Does a force have to be applied to a moving object to cause it to stop moving?

15. Which force is greater if other conditions are the same—rolling friction or sliding friction?

16. Someone who builds rockets and tries to make them faster, cheaper, and safer is called a rocket _____.

Creative Interpretation: Each image is worth 10 points.

17. Sometimes we have one set of limited facts and can come to different conclusions. Examine the following image and what "remains" of a painting. Then draw two possible reconstructions.

Remains of the original painting   Possible reconstruction #1   Possible reconstruction #2

| *Forces & Motion* | Quiz 2 | Scope: | Total score: | Name |
| Concepts & Comprehension | Level 1 | Chapters 5–8 | ____of 100 | |

Choose answers from these terms.

All the terms may not be used and some may be used more than once:

air	Bernoulli	buoyant	density	engineer
falls	friction	gravity	inertia	lift
mass	mountaintop	no	resistance	same
sea level	scientist	volume	yes	

Fill in the Blank: Each blank is worth 5 points.

1. All matter has a tendency to keep moving once it is in motion. This property is called _____.

2. Does air have weight?

3. Is air pressure greater at sea level or on a mountaintop?

4. Will a balloon get larger if the air pressure inside the balloon remains the same, but the air pressure outside becomes less?

5. Is there gravity on the moon?

6. A feather would fall a fast as a hammer on the moon, because there is no _____ around the moon.

7. What is the unbalanced force pulling down on a falling object?

8. Does air's resistance push up on a falling object?

9. Does air fill the space inside an "empty" cup?

10. Does friction push on a moving object opposite to the way the object is moving?

11. Archimedes used a method known as water displacement to measure what?

12. Archimedes discovered that there is a _____ force on objects floating in water that is equal to the weight of the water the object displaces.

13. Would a solid block of iron displace the same amount of water as a boat made from an equal block of iron?

14. If you know the volume and the weight of an object, you can calculate what?

15. What is the name of the force that pushes upward on an airplane wing?

16. Are the forces on a floating ship balanced if the buoyant force is equal to the weight of the ship?

17. A hammer and a feather fall at the _____ speed on the moon because there is no _____ around the moon. Thus, there is no air _____ on the feather as it _____.

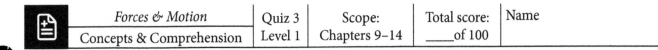

	Forces & Motion	Quiz 3	Scope:	Total score:	Name
	Concepts & Comprehension	Level 1	Chapters 9–14	____of 100	

Choose answers from these terms.

All the terms may not be used and some may be used more than once:

air	Archimedes	force	friction	gravity
greater	inertia	less	matter	moving
Newton	no	reaction	space	speed
torsion	yes			

Fill in the Blank: Each blank is worth 5 points.

1. When matter is at rest, it has a tendency to remain at rest. This property of matter is called

    _____.

2. Suppose a cream pie is sitting under the back window of a car that hits a brick wall at 30 miles per hour. Will the pie continue to sit under the back window when the car stops?

3. When an object is in motion, is a force needed to cause it to keep on moving?

4. Suppose you are sitting in a sled that is traveling down a snow-covered hill. Will your speed remain the same all the way down the hill?

5. According to Newton's second law of motion, the faster an object accelerates, the more _____ it will have.

6. As a toy car rolls down a ramp, what causes the car to travel faster and faster?

7. After a toy car has started to move, what is the main force that opposes its motion?

8. Will a Mexico cliff diver and someone who dives off the high dive at a swimming pool be traveling at the same speed when they hit the water?

9. What kind of force was used by early catapults to launch a missile?

10. What scientist wrote three laws of motion, a law of universal gravitation, invented calculus, and learned many new things about light?

11. Does the speed of the moon increase as it orbits the earth?

12. Does the size of the earth's orbit around the sun remain the same?

13. What force causes the moon to travel in an orbit around the earth instead of going straight?

14. Does the earth's gravitation attraction pull as hard on an object in space as it would at sea level?

15. Can a rocket go faster through the air or in space?

16. According to Newton's third law of motion, for every action, there is an equal and opposite
_____.

17. All _____ has a tendency to keep moving once it starts _____. If a toy car were pushed
across a smooth table, it would keep moving straight ahead at the same _____ until
_____ caused it to slow down and stop.

Forces & Motion | Quiz 4 | Scope: | Total score: | Name
Concepts & Comprehension | Level 1 | Chapters 15–20 | ____of 100 |

Choose answers from these terms.
All the terms may not be used and some may be used more than once:

axle	big	faster	fixed	harder	yes
higher	lower	moveable	small	wheel	no

Fill in the Blank: Each question is worth 10 points.

1. Tightrope walkers usually carry a long, heavy, curved pole as they walk across the rope. They use the pole to make their center of gravity _____.

2. Can you change the center of gravity of an object by adding weights to it?

3. To lift a heavy object with a lever, should you place the fulcrum far away from the object?

4. The advantage of using a fly swatter as a simple machine is to hit the fly _____.

5. The kind of pulley that makes it easier to lift a heavy weight is a _____ pulley.

6. Suppose someone is able to use a block and tackle to lift a 300-pound object using only 30 pounds of effort. Is the disadvantage of such a machine that the heavy object will move very slowly?

7. Is a wheel and axle one simple machine?

8. Some screwdrivers have big handles and some have small handles. Which would be easier to use if you need to turn a screw into a piece of wood?

9. Look at the picture of the screwdriver, a wheel and axle simple machine. The part labeled x is the _____.

10. Does sliding a box up a ramp take more effort than lifting the box straight up?

Forces & Motion	Test 1	Scope:	Total score:	Name
Concepts & Comprehension	Level 1	Chapters 1–20	____of 100	

Fill in the Blank Questions: (3 points each)

Choose answers from these terms.
All the terms may not be used and some may be used more than once:

air	Archimedes	axle	Bernoulli	big
buoyant	centimeters	density	engineer	faster
fixed	force	frame of reference	friction	gravity
greater	harder	heat	higher	increase
inertia	less	lift	lower	mass
mountaintop	moveable	Newton	no	north
reaction	reduce	rolling friction	scientist	scientist
sea level	second	sliding friction	small	south
space	streamlined	time	torsion	volume
wheel	yes			

1. In order to find something's speed, you need to divide the distance by the _____.

2. In order to tell how something is moving, you need a _____.

3. What force causes your tennis shoes to wear out?

4. What force helps you to walk without falling down?

5. Friction always produces what form of energy?

6. Ball bearings and roller bearings are used to _____ friction.

7. Is it possible for the same set of facts to be interpreted in more than one way?

8. Does a force have to be applied to a moving object to cause it to stop moving?

9. All matter has a tendency to keep moving once it is in motion. This property is called _____.

10. Does air have weight?

11. Is there gravity on the moon?

12. A feather would fall as fast as a hammer on the moon, because there is no _____ around the moon.

13. What is the unbalanced force pulling down on a falling object?

14. Someone who builds rockets and tries to make them faster, cheaper, and safer is called a rocket _____.

15. Archimedes discovered that there is a _____ force on objects floating in water that is equal to the weight of the water the object displaces.

16. If you know the volume and the weight of an object, you can calculate what?

17. What is the name of the force that pushes upward on an airplane wing?

18. _____'s Principle states that the faster a column of air moves, the less the internal pressure of the air will be.

19. When matter is at rest, it has a tendency to remain at rest. This property of matter is called _____.

20. When an object is in motion, is a force needed to cause it to keep on moving?

21. According to Newton's second law of motion, the faster an object accelerates, the more _____ it will have.

22. As a toy car rolls down a ramp, what causes the car to travel faster and faster?

23. Does the speed of the moon increase as it orbits the earth?

24. What force causes the moon to travel in an orbit around the earth instead of going straight?

25. Can a rocket go faster through the air or in space?

26. Can you change the center of gravity of an object by adding weights to it?

27. To lift a heavy object with a lever, should you place the fulcrum far away from the object?

28. The advantage of using a fly swatter as a simple machine is to hit the fly _____.

29. The kind of pulley that makes it easier to lift a heavy weight is a _____ pulley.

30. Do all simple machines have moving parts?

Essay: Each of the following questions is worth 5 points.

31. Describe the forces at work in this illustration:

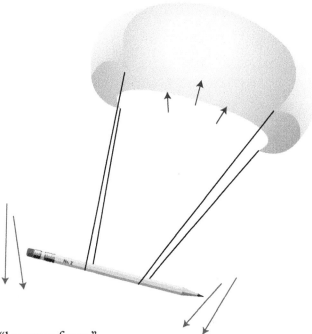

32. Label this illustration showing "weight force" and "buoyant force." Then explain how the ship stays afloat in the water because of these two forces.

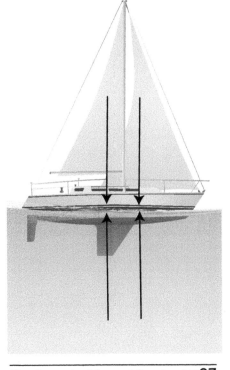

Multiple Choice: Please select the best answer. Each question is worth 5 points.

1. To calculate the speed of a moving object
   (a) add the distance and the time
   (b) multiply the distance and the time
   (c) divide the distance by the time

2. A line graph of a runner looks like this:
   What does this mean?
   (a) The runner slowed down at the end of the race.
   (b) The runner maintained the same speed throughout the race.
   (c) The runner kept going faster and faster.

3. Which of the following does not describe motion?
   (a) change of position   (b) speed   (c) weightlessness

4. Suppose you are on an airplane that is traveling 300 miles per hour. How many miles would you travel in 3 hours?
   (a) 100 miles   (b) 600 miles   (c) 900 miles

5. Which of the following indicates "two hundred meters per second"?
   (a) 200 m/s   (b) 200 mi/s   (c) 200 m.s.

6. A student ran 100 meters in 20 seconds. What is her speed?
   (a) 5   (b) 5 m   (c) 5 m/sec

7. Suppose you are riding a large float boat down a river that is moving south at 15 miles/hour. You are walking slowly in a northerly direction. To a friend on the riverbank watching you, you would appear to be moving:
   (a) north   (b) south   (c) not moving at all

8. Suppose you are looking at a series of pictures of several boats in a lake. What would you need in order to tell how the boats were moving?
   (a) a frame of reference
   (b) a comparison of any two boats
   (c) the position of the sun

9. Is it possible for the same set of facts to be interpreted in more than one way?
   (a) yes   (b) no

10. What force causes your tennis shoes to wear out?
    (a) gravity   (b) tension forces   (c) friction

11. Friction always produces what form of energy?
    (a) electricity   (b) light   (c) heat

12. Friction always pushes
    (a) in the same direction an object is moving
    (b) opposite to the way an object is moving
    (c) upward

13. If there was no friction between a car and the highway, could the car turn a curve?
    (a) yes   (b) no

14. What would not cause the amount of friction between two surfaces to increase?
    (a) rougher surfaces
    (b) smoother surfaces
    (c) faster motion between the surfaces

15. Would your pencil write if there was no friction between the pencil and the paper?
    (a) yes   (b) no

16. Which one of the following is a form of energy?
    (a) friction   (b) gravity   (c) heat

17. What would produce the most heat?
    (a) rubbing your hands together slowly
    (b) rubbing your hands together rapidly
    (c) there would be no difference

18. What causes a moving object to stop moving?
    (a) a moving object naturally stops moving
    (b) an unbalanced force
    (c) balanced forces

19. Does a force have to be applied to an object to cause it to start moving?
    (a) yes   (b) no

20. Which of the following would not affect the amount of friction produced by a moving object?
    (a) degree of smoothness
    (b) weight of the object
    (c) magnetic field

Multiple Choice: Please select the best answer. Each question is worth 5 points.

1. The upward and downward forces on an airplane wing (in flight) are balanced if the lift force is equal to:
   (a) the thrust forces   (b) the friction forces   (c) the weight of the airplane

2. Does air have weight?
   (a) yes   (b) no

3. About how much air pressure is there on every square inch at sea level?
   (a) 5 pounds   (b) 10 pounds   (c) 15 pounds

4. Is air pressure greater at sea level or on a mountaintop?
   (a) at sea level   (b) on a mountaintop   (c) there is no difference.

5. Suppose the air pressure inside a balloon remains the same, but the pressure outside becomes less. Assume the temperature remains the same. What will happen to the balloon?
   (a) It will get bigger.
   (b) It will get smaller.
   (c) It will stay the same size.

6. What forces act on falling objects?
   (a) downward pull of gravity and upward push of air resistance
   (b) downward pull of gravity only
   (c) downward pull of inertia

7. Is there gravity on the moon?
   (a) yes   (b) no

8. Why would a feather fall as fast as a hammer on the moon?
   (a) The moon's gravity is very strong.
   (b) There is no air around the moon.
   (c) There is no magnetic field around the moon.

9. A cup of water is covered by a card and turned upside-down. Why doesn't the card fall off?
   (a) Because of the water's inertia.
   (b) The air pressure on the card is greater than the weight of the water pushing on the card.
   (c) There's no logical explanation.

10. Racing cars, airplanes, rockets, and speed boats are all designed to have a special shape that reduces friction with air or water. What is this shape called?
    (a) streamlined   (b) polyphyletic   (c) polymorphic

11. The volume of an irregular solid can be found by what method?
    (a) weighing the object on balance scales
    (b) estimating its circumference and calculating the volume of a sphere
    (c) measuring the volume of water the object displaces

12. In order to calculate the density of an object, what two things must be known?
    (a) length and width
    (b) length and volume
    (c) volume and weight (mass)

13. What force pushes up on objects that are under water?
     (a) gravity   (b) buoyant force   (c) torsion force

14. Who discovered a way to tell how much force pushes up on objects that float (or sink) in a fluid?
     (a) Bernoulli   (b) Newton   (c) Archimedes

15. The buoyant force on an object that is floating in water is equal to:
     (a) the weight of the water the object displaces
     (b) 15 pounds per square inch
     (c) air pressure above the object

16. Which would displace the most water?
     (a) a solid block of iron
     (b) a boat made from the same block of iron
     (c) there would be no difference

17. Why does a heavy object feel lighter under water than it does out of water?
     (a) It is being pushed up by the force of lift.
     (b) It is being pushed up by buoyant force.

18. What is the name of the force that pushes upward on an airplane wing?
     (a) buoyant force   (b) lift   (c) magnetism

19. Where is the air moving faster around an airplane wing?
     (a) over the top of the wing
     (b) under the bottom of the wing
     (c) there is no difference

20. Bernoulli's principle says that the air pressure inside a column of air is less where the air:
     (a) is moving fastest   (b) is moving slowest   (c) is not moving at all.

Multiple Choice: Please select the best answer. Each question is worth 4 points.

1. Objects in motion tend to stay in motion at the same speed and direction. Objects at rest tend to remain at rest. This property of matter is called what?
    (a) momentum   (b) density   (c) inertia

2. Suppose a penny is placed on top of a toy car as the car rolls down a ramp and hits a barrier. What would the penny do when the car stopped?
    (a) The penny would remain on the toy car.
    (b) The penny would move straight backward.
    (c) The penny would move straight ahead.

3. Suppose an object in a car is not tied down when the car hits a brick wall while moving at 30 miles per hour. What will the object do at first?
    (a) It will stop as soon as the car stops.
    (b) It will continue to move forward at 30 miles per hour.
    (c) It will move backward at 30 miles per hour.

4. When an object is in motion, is a force needed to cause it to keep on moving?
    (a) yes   (b) no

5. A famous magician trick is to make a tall stack of breakable glassware on a table and then pull the tablecloth out from under the glass. If the tablecloth is made of a smooth fabric with no hem or rough edges, the glassware will barely move. This is because:
    (a) Objects at rest tend to remain at rest.
    (b) Friction is a big force.
    (c) For every action, there is an equal and opposite reaction.

6. Which sled would travel farther coming off a snow-covered hill?
    (a) a sled at the top of the hill
    (b) a sled half-way up the hill
    (c) they would both travel the same distance

7. If you slide down a snow-covered hill on a sled,
    (a) your speed will stay the same all the way down.
    (b) you will travel faster and faster all the way down the hill.
    (c) your speed will become slower as you travel down the hill.

8. According to Newton's second law of motion, the faster a falling object accelerates, the more _____ it will have.
    (a) gravitational pull   (b) force   (c) mass

9. As a toy car rolls down a ramp, what causes the car to travel faster and faster?
    (a) inertia   (b) the earth's gravitational pull   (c) air resistance

10. After a toy car has started to move, what is the main force that causes it to stop moving?
    (a) friction
    (b) magnetic forces
    (c) it is natural for moving objects to stop moving

11. When an object keeps on changing its speed by going faster and faster, it is said to be
    (a) streamlining   (b) pressurizing   (c) accelerating

12. Compare the speed of a Mexico cliff diver with a person who dives off the high dive at a swimming pool.
    (a) The cliff diver is traveling much faster than the diver in a swimming pool.
    (b) The diver in a swimming pool is traveling much faster than the cliff diver.
    (c) Both divers are traveling at the same speed when they hit the water.

13. What kind of powerful war machines were designed by the Greeks and Romans before the time of Christ?
    (a) rockets   (b) airplanes   (c) catapults

14. What will happen to a missile's speed after it has been launched into the air?
    (a) It will continue to go faster and faster.
    (b) It will continue to travel at the same speed until other forces slow it down.

15. What kind of force is the result of twisting?
    (a) torsion   (b) tension   (c) hydraulic

16. Planes taking off from an aircraft carrier (ship) only have a few seconds to accelerate to the necessary flying speed. How do they get enough speed to fly?
    (a) They have powerful engines that let them move very fast.
    (b) They are launched by a catapult on the ship.

17. What scientist wrote three laws of motion, a law of universal gravitation, invented calculus, and learned many new things about light?
    (a) Archimedes   (b) Bernoulli   (c) Newton

18. What would happen to an object that was moving at 20 miles per hour if there were no unbalanced forces acting on it?
    (a) It would quickly slow down.
    (b) It would speed up.
    (c) It would continue to move straight ahead at 20 miles per hour indefinitely.

19. You made a model that represented the earth and the moon using a stopper that you could whirl around in a circular motion. In this model, the weight and the string represented
    (a) inertia   (b) gravitational pull of the earth   (c) friction

20. The size of the moon's orbit around the earth:
    (a) stays the same   (b) gets bigger   (c) gets smaller

21. As the earth orbits the sun, its speed:
    (a) remains the same   (b) increases   (c) decreases

22. What force causes the moon to travel in an orbit around the earth instead of going straight?
    (a) gravity   (b) friction   (c) buoyancy

23. Does the earth's gravitational attraction pull as hard on an object in space as it would at sea level?
    (a) yes   (b) no

24. Which is not true?
    (a) Mass doesn't change because it is a measure of how much matter there is.
    (b) Weight may change because it is determined by how hard gravity pulls on an object.
    (c) Both the mass and the weight of an object would change if it went from the earth to the moon.

25. In an activity, two students on skateboards throw a 5-pound bag of flour. One student weighs 80 pounds and one student weighs 120 pounds. What will happen to the students?
    (a) The 80-pound student will move back faster (and farther) than the other student.
    (b) The 120-pound student will move back faster (and farther) than the other student.
    (c) Both students will move back at the same speed.

Multiple Choice: Please select the best answer. Each question is worth 4 points.

1. Can you change the center of gravity of an object by adding weights to it?
   (a) yes   (b) no

2. Is the center of gravity of a 20-story building at the 10th story?
   (a) yes   (b) no

3. Engineers try to design vehicles that don't turn over easily. They do this by:
   (a) making the center of gravity as low as practical
   (b) making the center of gravity as high as practical
   (c) making the center of gravity in the exact center of the vehicle

4. What is the effect of the extra weight of a long, curved pole held by a tightrope walker?
   (a) It lowers the entertainer's center of gravity.
   (b) It raises the entertainer's center of gravity.
   (c) It makes it harder for the entertainer to maintain balance.

5. A spinning top has rotational:
   (a) inertia   (b) magnetism   (c) buoyancy

6. To lift a heavy object with a lever, where would you place the fulcrum?
   (a) close to the object   (b) far away from the object   (c) next to your hand

7. A fly swatter is an example of what kind of simple machine?
   (a) a wheel and axle   (b) an inclined plane   (c) a lever

8. What is the advantage of using a fly swatter as a simple machine?
   (a) to hit the fly faster
   (b) to hit the fly harder
   (c) to move your hand in a different direction

9. A crowbar is a simple machine that allows you to move a heavy object with a small force. A disadvantage of using a crow bar is that:
   (a) the object moves faster than you hand
   (b) the object only moves a short distance
   (c) there is an excessive amount of friction when the object is being moved

10. Which of the following levers lets you move an object much faster than your hand is moving?
    (a) golf club   (c) car jack   (c) crowbar

11. If a lever's effort arm is 2 meters long and its load arm is 1 meter long, what is the mechanical advantage of the lever?
    (a) 1   (b) 2   (c) 3

12. The kind of pulley that makes it easier to lift a heavy weight is a:
    (a) fixed pulley   (b) moveable pulley

13. What kind of pulley lets you pull down on a rope in order to raise a flag up a flagpole?
    (a) a fixed pulley   (b) a moveable pulley

14. Machines have a mechanical advantage that tells you how many times your effort force should be multiplied by the machine. What force will actually cause your effort force to be greater than this number?
   (a) gravity   (b) friction   (c) tension

15. Suppose someone is able to use a block and tackle system of pulleys to lift a 300-pound weight using only 30 pounds of effort force. Will the 300-pound weight move very fast or very slowly?
   (a) very fast   (b) very slowly

16. Is a single pulley a simple machine?
   (a) yes   (b) no

17. When you use a single moveable pulley to lift a weight, which moves the greater distance?
   (a) the weight   (b) your hand   (c) they will move equal distances.

18. Is a wheel and axle a simple machine?
   (a) yes   (b) no

19. Which kind of windlass would allow you to move a bucket of water easier?
   (a) one with a big crank (the wheel) and a small drum (the axle)
   (b) one with a big crank (the wheel) and a big drum (the axle)

20. What is the advantage of a screwdriver with a big handle compared to a screwdriver with a small handle?
   (a) It would be easier for the one with the big handle to turn a screw into a piece of wood.
   (b) There would be no difference in using a screwdriver with a big handle and one with a small handle.

21. Skates with wheels that turn around a stationary axle are not the same as a wheel and axle machine. Why not?
   (a) In a wheel and axle machine, only the axle turns.
   (b) In a wheel and axle machine, only the wheel turns.
   (c) In a wheel and axle machine, the wheel and axle are connected and turn together.

22. Look at the drawing of the screwdriver.
   The part labeled x is the
   (a) wheel   (b) axle

23. Which would be harder (take the most effort force)?
   (a) sliding a heavy box up a loading ramp
   (b) lifting the box straight up

24. Which would be harder (take the most effort force)?
   (a) sliding the box up a long loading ramp
   (b) sliding the box up a short loading ramp
   (c) there would be no difference

25. Do all simple machines have moving parts?
   (a) yes   (b) no

Multiple Choice Questions: (2 points each)

1. To calculate the speed of a moving object:
    (a) add the distance and the time
    (b) multiply the distance and the time
    (c) divide the distance by the time

2. Which of the following does not describe motion?
    (a) change of position   (b) speed   (c) weightlessness

3. Suppose you are on an airplane that is traveling 300 miles per hour. How many miles would you travel in 3 hours?
    (a) 100 miles   (b) 600 miles   (c) 900 miles

4. Which of the following indicates "two hundred meters per second"?
    (a) 200 m/s   (b) 200 mi/s   (c) 200 m.s.

5. Is it possible for the same set of facts to be interpreted in more than one way?
    (a) yes   (b) no

6. What force causes your tennis shoes to wear out?
    (a) gravity   (b) tension forces   (c) friction

7. Friction always produces what form of energy?
    (a) electricity   (b) light   (c) heat

8. If there was no friction between a car and the highway, could the car turn a curve?
    (a) yes   (b) no

9. Does a force have to be applied to an object to cause it to start moving?
    (a) yes   (b) no

10. What is the purpose of having ball bearings in skates?
    (a) to reduce friction   (b) to increase friction   (c) to reduce static electricity

11. The tendency of all matter to resist starting to move (or to keep moving once it is in motion) is known as
    (a) inertia   (b) momentum   (c) torsion

12. The upward and downward forces on an airplane wing (in flight) are balanced if the lift force is equal to the: (a) thrust forces, (b) friction forces, (c) weight of the airplane.

13. Does air have weight? (a) yes (b) no

14. Suppose the air pressure inside a balloon remains the same, but the pressure outside becomes less. Assume the temperature remains the same. What will happen to the balloon?
    (a) It will get bigger.   (b) It will get smaller.   (c) It will stay the same size.

15. What forces act on falling objects?
    (a) downward pull of gravity and upward push of air resistance
    (b) downward pull of gravity only
    (c) downward pull of inertia

16. Is there gravity on the moon?
    (a) yes   (b) no

17. Why would a feather fall as fast as a hammer on the moon?
    (a) The moon's gravity is very strong.
    (b) There is no air around the moon.
    (c) There is no magnetic field around the moon.

18. Racing cars, airplanes, rockets, and speed boats are all designed to have a special shape that reduces friction with air or water. What is this shape called?
    (a) streamlined   (b) polyphyletic   (c) polymorphic

19. In order to calculate the density of an object, what two things must be known?
    (a) length and width   (b) length and volume   (c) volume and weight (mass)

20. What force pushes up on objects that are under water?
    (a) gravity   (b) buoyant force   (c) torsion force

21. Why does a heavy object feel lighter under water than it does out of water?
    (a) It is being pushed up by the force of lift.
    (b) It is being pushed up by buoyant force.

22. The principle that explains why an airplane overcomes the downward pull of gravity is known as:
    (a) Archimedes' principle (b) Newton's third law of motion (c) Bernoulli's principle

23. Objects in motion tend to stay in motion at the same speed and direction. Objects at rest tend to remain at rest. This property of matter is called what?
    (a) momentum   (b) density   (c) inertia

24. When an object is in motion, is a force needed to cause it to keep on moving?
    (a) yes   (b) no

25. Which sled would travel farther coming off a snow-covered hill?
    (a) a sled at the top of the hill
    (b) a sled halfway up the hill
    (c) they would both travel the same distance

26. According to Newton's second law of motion, the faster a falling object accelerates, the more
    _____ it will have.
    (a) gravitational pull   (b) force   (c) mass

27. As a toy car rolls down a ramp, what causes the car to travel faster and faster?
    (a) inertia   (b) the earth's gravitational pull   (c) air resistance

28. After a toy car has started to move, what is the main force that causes it to stop moving?
    (a) friction
    (b) magnetic forces
    (c) it is natural for moving objects to stop moving

29. When an object keeps on changing its speed by going faster and faster, it is said to be
    (a) streamlining   (b) pressurizing   (c) accelerating

30. What kind of powerful war machines were designed by the Greeks and Romans before the time of Christ?
    (a) rockets   (b) airplanes   (c) catapults

31. What kind of force is the result of twisting?
    (a) torsion   (b) tension   (c) hydraulic

32. What scientist wrote three laws of motion and a law of universal gravitation, invented calculus, and learned many new things about light?
    (a) Archimedes   (b) Bernoulli   (c) Newton

33. The size of the moon's orbit around the earth:
    (a) stays the same   (b) gets bigger   (c) gets smaller

34. As the earth orbits the sun, its speed
    (a) remains the same   (b) increases   (c) decreases

35. What force causes the moon to travel in an orbit around the earth instead of going straight?
    (a) gravity   (b) friction   (c) bouyancy

36. In a moving rocket, there is an "action" and a "reaction." The "reaction" occurs when
    (a) the rocket pushes gases out one end of the rocket.
    (b) the gases push back on the rocket.

37. Can a rocket go faster through the air or in space?
    (a) air   (b) space

38. Which statement is true according to Newton's third law of motion?
    (a) The action is greater than the reaction.
    (b) The reaction is greater than the action.
    (c) The action and the reaction are the same.

39. Suppose an astronaut was outside a space ship (in a weightless condition), and he/she pushed against the side of the space ship. What would happen?
    (a) The astronaut would move toward the space ship.
    (b) The astronaut would move away from the space ship.

40. Can you change the center of gravity of an object by adding weights to it?
    (a) yes   (b) no

41. Engineers try to design vehicles that don't turn over easily. They do this by
    (a) making the center of gravity as low as practical.
    (b) making the center of gravity as high as practical.
    (c) making the center of gravity in the exact center of the vehicle.

42. A spinning top has rotational _____.
    (a) inertia   (b) magnetism   (c) buoyancy

43. To lift a heavy object with a lever, where would you place the fulcrum?
    (a) close to the object   (b) far away from the object (c) next to your hand

44. A fly swatter is an example of what kind of simple machine?
    (a) a wheel and axle   (b) an inclined plane   (c) a lever

45. What is the advantage of using a fly swatter as a simple machine?
    (a) to hit the fly faster
    (b) to hit the fly harder
    (c) to move you hand in a different direction

46. The kind of pulley that makes it easier to lift a heavy weight is a
    (a) fixed pulley   (b) moveable pulley

47. What kind of pulley lets you pull down on a rope in order to raise a flag up a flagpole?
    (a) a fixed pulley   (b) a moveable pulley

48. What is the advantage of a screwdriver with a big handle compared to a screwdriver with a small handle?
    (a) It would be easier for the one with the big handle to turn a screw into a piece of wood.
    (b) There would be no difference in using a screwdriver with a big handle and one with a small handle.

49. Which would be harder (take the most effort force)?
    (a) sliding the box up a long loading ramp
    (b) sliding the box up a short loading ramp
    (c) there would be no difference

50. Do all simple machines have moving parts?
    (a) yes   (b) no

Bonus Question: 5 Points

Skates with wheels that turn around a stationary axle are not the same as a wheel and axle machine.
Why not?

    (a) In a wheel and axle machine, only the axle turns.
    (b) In a wheel and axle machine, only the wheel turns.
    (c) In a wheel and axle machine, the wheel and axle are connected and turn together.

Quizzes and Test

for Use with

Earth

Testing:

This series is appropriate for both upper elementary and junior high students. Because of this, we have included quizzes and tests in two different levels which you can choose from based on your child's abilities and understanding of the concepts in the course.

Level 1: suggested for younger ages or those who struggle with application of the concepts beyond just definitions and basic concepts

Level 2: suggested for older ages or those who can both grasp the scientific concepts and apply them consistently

Permission to Copy

Fill in the Blank Questions: (4 points each)

Choose answers from these terms.
All the terms may not be used and some may be used more than once:

building materials	circle of lava	coal	color
concave-type	convection-type	crust	earthquake zone
earthquakes	elastic rebound theory	engineering	ethanol
five	gas	general position sector	global positioning system
Greenwich (or	height	helium	Himalayan Mountains
Prime) Meridian	hydrogen	iron and nickel	latitudinal line(s)
less	longitudinal line(s)	magma	mineral deposits
Mount Rushmore	New Madrid Fault	nickel and iron	oil
on or near the equator	ores	oxygen	Pangaea
Ring of Fire	San Andreas Fault	seafloor spreading	steel
subduction	surface	tsunamis	twelve
twenty-four (24)	under land	under ocean	water
weight			

1. The International Date Line is a _____.

2. What is the name of the starting longitudinal line that is designated as 0°? _____

3. Which lines go from the North Pole to the South Pole? _____

4. Which lines circle the earth and are parallel to the equator? _____

5. What part of the earth doesn't have four seasons? _____

6. Into how many times zones is the earth divided? _____

7. What do the letters "GPS" mean when referring to a GPS device?

8. In which layer of the earth are solid rocks found that are not extremely hot? _____

9. Most scientists believe the core of the earth is made of what elements? _____

10. Is the earth's crust thicker under the continents or under the oceans? _____

11. Some geologists believe there are _____currents in parts of the earth's mantle and core.

12. What are three natural resources found in the earth's crust?
    a.
    b.
    c.

13. Is the Mid-Atlantic Ridge a region of subduction or seafloor spreading?_____

14. _____ occur frequently along or near the San Andreas Fault in western California.

15. The _____ are thought to have formed when two crustal plates collided into each other, but neither plate slid under the other one.

16. _____ may have been the original land mass that broke apart and formed today's continents.

17. According to the _____, rocks break loose from a position of tension — and suddenly surge forward.

18. An earthquake begins as locked-up sections of rocks break free, creating more or less tension on the rocks? _____

19. Crustal plate movement is occurring along the fault line known as the _____ as the Pacific Plate is slowly moving past the North American Plate.

20. One of the main earthquake belts in the earth is known as the "_____."

Short Answer: Each question is worth 5 points.

21. What can a GPS device in an automobile do?

22. Where are earthquakes most prone to occur?

23. What is happening where the Pacific Plate and the North American Plate meet along the coast of California?

24. The three main states of matter are solid, liquid, and gas. What is meant by a plastic state?

Bonus Question: (worth 5 points)

25. Sometimes buildings fall in during an earthquake. What are two things that play a big role in how well a building can withstand an earthquake?

Fill in the Blank Questions: (6 points each)

Choose answers from these terms.
All the terms may not be used and some may be used more than once:

Badlands (South Dakota) crevices   every day   every three months
extrusive   faults   Galapagos   glaciers
Grand Canyon   Great Prairie   Hawaii   intrusive
lava   lava flows   Mount Rushmore   nine
no   once a week   Richter scale   seismograph
sonar   Surtsey   three   tornadoes
volcanic eruptions   volcanoes   water   wind
yearly   yes

1. The strength of an earthquake is reported as a number from 0 to 10, known as the _____.

2. What is the name of the instrument that is used to study and identify earthquakes?

   _____

3. How often do earthquakes occur throughout the world?

4. Which earthquake is more powerful — one that measures 3 on the Richter scale or one that measures 9 on the Richter scale? _____

5. When magma reaches the surface of the earth, it is called_____.

6. Were there warning signs that Mount St. Helens was about to erupt before May 18, 1980? _____

7. _____, like earthquakes, tend to be found where large crustal plates meet. Sometimes they are found in areas known as "hot spots."

8. Are volcanic eruptions always violent and explosive?

9. Give an example of a place in the United States where miles of flat, level layers of strata can be seen.

   _____

10. Name at least three ways in which sedimentary layers can be laid down in nature.
    a.
    b.
    c.

11. Which of the following processes is most likely to produce flat level layers of sediment — glaciers, wind, water, or volcanic eruptions?_____

12. What do we call breaks and cracks in large rock formations when rock on one side of the crack has slipped and moved? _____

13. _____ rocks form when magma reaches the surface of the earth and hardens there.

14. _____ rocks are rocks that cool and harden from hot magma below the surface of the earth.

15. Are dikes and sills intrusive or extrusive rocks? _____

Short Answer: Each question is worth 5 points.

16. What are some ways the government of a country could reduce the deaths and damage caused by an earthquake?

17. Where are the two main earthquake belts in the earth?

Bonus Question: (worth 5 points)

18. This island first appeared in 1963. _____.

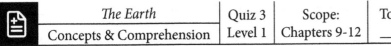

	The Earth	Quiz 3	Scope:	Total score:	Name
	Concepts & Comprehension	Level 1	Chapters 9-12	____of 100	

Fill in the Blank Questions: (6 points each)

Choose answers from these terms.
All the terms may not be used and some may be used more than once:

air and water	amphiboles	animals	calcite
calcium carbonate	clay minerals	cleavage	color
composite	crystal form	defined shape	density
feldspar(s)	flame	flat surface	fluorescence
gradual	hardness	heat	igneous
large	limestone	luster	magnetic properties
marble	metamorphic	mica(s)	minerals
no	olivines	plants	pressure
pyroxenes	quartz	radioactivity	rocks
rocks and dirt	sea level	sedimentary	silicate
silver	small	steep	streak
sulphur	texture	topographic	yes

1. Do contour lines overlap each other? _____

2. Contour lines on a map that are close together indicate a _____ incline.

3. Contour lines that are spaced farther apart indicate a _____ incline.

4. A _____ map shows contour lines, as well as other details of specific features in an area.

5. Do all crystals have the same shape? _____

6. When crystals form from conditions in which they grow rapidly, they tend to be _____.

7. Can the shape of crystals be used to help identify kinds of crystals?

8. Crystals found in nature are called _____.

9. Name the minerals found in granite.
    a.
    b.
    c.

10. _____ minerals make up about 95 percent of the minerals in the earth's crust.

11. Name at least four of eight minerals that make up most of the rocks in the earth's crust.
    a.
    b.
    c.
    d.

12. What is the name of the chemical in limestone? _____

13. Would you be more likely to find fossils in limestone, marble, or granite rocks? _____

14. There are three main groups of rocks — sedimentary, metamorphic, and igneous. In which group would limestone be classified?

15. What two conditions are generally necessary in order for metamorphic rocks to form?
    a.
    b.

Short Answer: Each question is worth 5 points.

16. Limestone often contains the fossilized remains of what kinds of things?

17. Whether a contour map shows the depth of the ocean bottom or the elevation of dry land, what is the starting point for measuring these distances? _____

Bonus Questions: (worth 5 points)

18. Why is granite classified as a rock instead of a mineral?

19. What are two differences in a crystal and a rock?

20. Name at least four of eight tests that can be done to identify minerals.
    a.
    b.
    c.
    d.

Fill in the Blank Questions: (6 points each)

Choose answers from these terms.
All the terms may not be used and some may be used more than once:

chemical weathering	clay	conglomerate	contracts
delta	deposition	dust	erosion
expands	glaciers	hills/bare ground	igneous
intrusive igneous	iron oxide	large/heavy	limestone
metamorphic	oxbow lake	parking lots/wooded areas	pebbles
physical weathering	sandbar	sandstone	sedimentary
sedimentation	shale	sharp/jagged	small/lighter
smooth/rounded	stream	topsoil	volcanic

1. Name the three big groups of rocks.
   a.
   b.
   c.

2. Which kind of rocks are formed when igneous rocks or sedimentary rocks are subjected to high temperatures and/or pressure? _____

3. Which kind of rock forms from the cooling of magma underground before it reaches the surface of the earth? _____

4. Name three sedimentary rocks.
   a.
   b.
   c.

5. Basalt, granite, and obsidian are examples of what kind of rock? _____

6. What term refers to materials, such as stones, sand, and silt, that are deposited by water? _____

7. What is another name for the process of sedimentation? _____

8. Is the settling rate faster for large/heavy materials or for small/lighter materials? _____

9. The Mississippi River carries tons of sediment into the Gulf of Mexico every year. What is the area called where the river and the gulf meet? _____

10. What is most likely to leave deposits of large boulders — water, wind, or glaciers? _____

11. Name two places where erosion occurs at a faster rate.
    a.   b.

12. When the curves of a meandering river meet and produce a new pathway for the river, the old river curve will remain and is called an _____.

13. Suppose a rock breaks into several small pieces after a big rock falls on it. Is this an example of physical weathering or chemical weathering? _____

14. Rocks sometimes break during freezing weather after water gets into cracks in the rocks and _____.

15. What kind of shape would you expect newly broken rocks to have? _____

Short Answer: Each question is worth 5 points.

16. It may take hundreds of years to produce a good, thick layer of _____, yet it can be lost in a few years by erosion.

17. _____is a process that leaves deposits of sediment behind.

Bonus Questions: (worth 5 points)

18. Iron is a strong, dark-colored, shiny metal; _____is a reddish-colored, dull material that crumbles easily.

19. Marble is a metamorphic rock that is made from what sedimentary rock under conditions of great heat and pressure?

20. Define the word "erosion."

Fill in the Blank Questions: (6 points each)

Choose answers from these terms.
All the terms may not be used and some may be used more than once:

aquifer	cistern	cold	continental
deep below	dirt	dirty	drought
dry	earthquake	flood	glaciers
gravel	holes	hot	humus
maybe	mountain	no	on the surface of
pores	porosity	puddle	runoff
saturated	stalactites	stalagmites	solid rock
spongy	valley	water	water table
wind	yes		

1. When farmers dig wells in order to get water to drink, these wells must be deep enough to reach the_____, the place where water has saturated a layer of rocks, sand, and other soil material.

2. When a water well goes _____, the water table might have become lower due to insufficient rain.

3. The spaces between rocks, sand, or other earth materials are called what? _____

4. The amount of spaces between rocks, sand, or other earth materials compared to the total volume of the material is known as what? _____

5. A rock formation with enough porosity to hold water and allow water to be moved in and out of the rock is called a water _____.

6. Most oil and gas deposits are _____ the ground.

7. When water has filled all the pores in a porous underground area, the area is said to be _____ with water.

8. When there is rain, some soaks into the ground, and that which does not is called _____.

9. As the water drops from the ceiling of a cave, small amounts of calcium carbonate are deposited. These deposits form slowly from the ceiling and are called _____.
Calcium carbonate formations in caves that rise from the floor are called _____.

10. Is the water that erupts from a geyser in Yellowstone National Park hot or cold?_____

11. A sinkhole is most likely to fall when there is a _____.

12. Till deposits and moraines are laid down by _____.

13. There are two kinds of glaciers: _____ and _____ glaciers.

14. _____ makes up a part of the top layer of soil and is made of decayed, weathered tree leaves, limbs, or other once-living things.

15. Can adding a bag of fertilizer to a field (where topsoil has been lost) make it as good as it was before the topsoil was lost? _____.

Short Answer: Each question is worth 5 points.

16. What determines how deep water soaks into the ground? _____
_____

17. Give an example of how underground rocks can be weathered physically. _____
_____

Bonus Questions: (worth 5 points)

18. Give an example of how underground rocks can be weathered chemically. _____
_____

19. Why is there a great need for people all over the world to find drinking water from clean wells rather than getting their drinking water from streams and shallow, polluted wells? _____
_____

The Earth	Test 1	Scope:	Total score:	Name
Concepts & Comprehension	Level 1	Chapters 1–20	____of 100	

Fill in the Blank Questions: (3 points each)

Choose answers from these terms.
All the terms may not be used and some may be used more than once:

amphiboles	aquifer	calcite	chemical weathering
clay minerals	conglomerate	continental/valley	deposition
earthquakes	elastic rebound	expands	extrusive
faults	feldspars	gradual	Greenwich Meridian
heat/pressure	heat/time	humid	humus
igneous	intrusive	iron/nickel	iron oxide
latitudinal	lava	limestone	longitudinal
metamorphic	maybe	micas	minerals
natural gas	no	oil	olivines
Pangaea	physical weathering	porosity	Prime Meridian
pyroxenes	quartz	Richter scale	Ring of Fire
runoff	sandstone	sediment	sedimentary
seismograph	shale	steep	time/pressure
topographic	tornadoes	valley/mountain	water
water table	wetlands	yes	

1. What is the name of the starting longitudinal line that is designated as 0°? _____

2. Which lines go from the North Pole to the South Pole? _____ lines

3. Which lines circle the earth and are parallel to the equator? _____ lines

4. Most scientists believe the core of the earth is made of what elements? _____

5. _____may have been the original land mass that broke apart and formed today's continents.

6. According to the _____theory, rocks break loose from a position of tension — and suddenly surge forward.

7. One of the main earthquake belts in the earth is known as the "_____."

8. The strength of an earthquake is reported as a number from 0 to 10, known as the _____.

9. What is the name of the instrument that is used to study and identify earthquakes?

_____

10. When magma reaches the surface of the earth, it is called_____.

11. _____ rocks form when magma reaches the surface of the earth and hardens there.

12. _____ rocks are rocks that cool and harden from hot magma below the surface of the earth.

13. Do contour lines on a topographic map overlap each other? _____

14. Contour lines on a map that are close together indicate a _____ incline.

15. A _____ map shows contour lines, as well as other details of specific features in an area.

16. Can the shape of crystals be used to help identify kinds of crystals?

17. Crystals found in nature are called _____.

18. Name at least four of eight minerals that make up most of the rocks in the earth's crust.

    a.

    b.

    c.

    d.

19. What two conditions are generally necessary in order for metamorphic rocks to form?

    a.

    b.

20. Name the three big groups of rocks.

    a.

    b.

    c.

21. Name three sedimentary rocks.

    a.

    b.

    c.

22. What term refers to materials such as stones, sand, and silt, that are deposited by water? _____
_____

23. What is another name for the process of sedimentation? _____

24. Suppose a rock breaks into several small pieces after a big rock falls on it. Is this an example of physical weathering or chemical weathering? _____

25. When farmers dig wells in order to get water to drink, these wells must be deep enough to reach the_____, the place where water has saturated a layer of rocks, sand, and other soil material.

26. The amount of spaces between rocks, sand, or other earth materials compared to the total volume of the material is known as what? _____

27. A rock formation with enough porosity to hold water and allow water to be moved in and out of the rock is called a water _____.

28. When there is rain, some soaks into the ground, and that which does not is called _____.

29. There are two kinds of glaciers: _____ and _____ glaciers.

30. _____makes up a part of the top layer of soil and is made of decayed, weathered tree leaves, limbs, or other once-living things.

Short Answer: Each question is worth 5 points

31. What are two differences in a crystal and a rock?

a.

b.

32. Why is there a great need for people all over the world to find drinking water from clean wells rather than getting their drinking water from streams and shallow, polluted wells?

Short Answer Questions: (5 points each)

1. Is the International Date Line a longitudinal line or a latitudinal line?

2. What is the name of the starting longitudinal line that is designated as 0°?

3. Which lines go from the North Pole to the South Pole?

4. Which lines circle the earth and are parallel to the equator?

5. What part of the earth doesn't have four seasons?

6. What is a GPS device? What can a GPS device in an automobile do?

7. The materials in the lower part of the mantle are extremely hot, but they are thought to be in a plastic state in places. Under what conditions does this plastic behave like a solid?

8. Under what conditions does material from the mantle rise up into the crust or even to the surface of the earth?

9. The three main states of matter are solid, liquid, and gas. What is meant by a plastic state?

10. Some geologists believe there are convection-type currents in parts of the earth's mantle and core. What are some ways these currents might affect the earth?

11. What are some of the natural resources found in the earth's crust?

12. What is happening where the Pacific Plate and the North American Plate meet along the coast of California?

13. Where is the Mid-Atlantic Ridge?

14. Is the Mid-Atlantic Ridge a region of subduction or seafloor spreading?

15. The Himalayan Mountains are thought to have formed when two crustal plates did what?

16. What do geologists believe about a land mass known as Pangaea?

17. According to the elastic rebound theory, what often causes the ground to shake during an earthquake?

18. Where are earthquakes most prone to occur?

19. What kind of crustal plate movement is occurring along the San Andreas Fault?

20. Is the New Madrid fault system located at the boundary of two major crustal plates?

Short Answer Questions: (5 points each)

1. The strength of an earthquake is reported as a number from 0 to 10. What is this scale of numbers called?

2. What is the name of the instrument that is used to study and identify earthquakes?

3. About how often do earthquakes occur throughout the world — every day or about once a month?

4. Which earthquake is more powerful — one that measures 3 on the Richter scale or one that measures 9 on the Richter scale?

5. What is the advantage of designing a building that moves slightly on its foundation?

6. What are some of the major earthquakes that have struck the United States?

7. Where are the two main earthquake belts on the earth?

8. When magma reaches the surface of the earth, what is it called?

9. What has probably happened if a volcano suddenly erupts and forcefully throws out volcanic materials and rocks?

10. Where are some of the most likely places to find volcanoes in the world?

11. When magma from the mantle is pushed up through cracks in the crust, does it always reach the surface of the earth?

12. Give an example of a place in the United States where miles of flat, level layers of strata can be seen.

13. Which of the following processes is most likely to produce flat, level layers of sediment — glaciers, wind, water, or volcanic eruptions?

14. What are some things that can change flat, horizontal sedimentary layers in nature after they have been laid down?

15. What do we call breaks and cracks in large rock formations when rock on one side of the crack has slipped and moved?

16. How does magma travel from the mantle through the crust of the earth?

17. When magma reaches the surface of the earth and hardens there, what kind of rock forms?

18. What are intrusive rocks?

19. Suppose you find an intrusive rock that has passed through several layers of stratified rocks. Which would be older — the intrusive rock or the stratified layers of rocks?

20. How does the cooling rate affect the size of mineral crystals that form in intrusive magma?

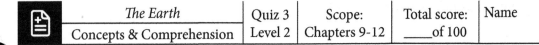

	The Earth	Quiz 3	Scope:	Total score:	Name
	Concepts & Comprehension	Level 2	Chapters 9-12	____of 100	

Short Answer Questions: (5 points each)

1. If "990 feet" is written next to a certain contour line on a map, what does that mean?

2. Do contour lines overlap each other?

3. What kind of map shows contour lines, as well as other details of specific features in an area?

4. Is the increase or decrease in height between any two contour lines on a map the same?

5. Whether a contour map shows the depth of the ocean bottom or the elevation of dry land, what is the starting point for measuring these distances?

6. Do all crystals have the same shape?

7. How can you tell the difference in a crystal and a rock?

8. Can the shape of crystals be used to help identify kinds of crystals?

9. What are crystalline structures that are found in nature called?

10. Do large alum crystals have the same shape as small alum crystals?

11. When crystals are found in nature, what are they called?

12. Why is granite classified as a rock instead of a mineral?

13. Are minerals and crystals pure substances or mixtures of chemicals?

14. Name at least eight tests that can be done to identify minerals.

15. What kinds of minerals make up about 95 percent of the minerals in the earth's crust?

16. What are three tests or observations that can help you identify limestone?

17. What is the name of the chemical in limestone?

18. There are three main groups of rocks — sedimentary, metamorphic, and igneous. In which group would limestone be classified? In which group would marble be classified?

19. What conditions are generally necessary for metamorphic rocks to form?

20. Limestone has many characteristics that make it a good building material. Why do builders often cover it with a sealant?

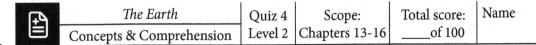

The Earth	Quiz 4	Scope:	Total score:	Name
Concepts & Comprehension	Level 2	Chapters 13-16	____ of 100	

Short Answer Questions: (5 points each)

1. Explain how to do a streak test on a rock.

2. Name the three big groups of rocks.

3. Which group of rocks contains fossils?

4. Which kind of rocks are formed when igneous rocks or sedimentary rocks are subjected to high temperatures and/or pressure?

5. Name three or four sedimentary rocks.

6. Basalt, granite, and obsidian are examples of what kind of rock?

7. Marble is a metamorphic rock that is made from what sedimentary rock under conditions of great heat and pressure?

8. What is the process of depositing sediment out of water?

9. What are two other things that can deposit sediments?

10. Is the settling rate faster for large/heavy materials or for small/lighter materials?

11. Where are sand bars likely to be found — on the inside of a river curve where the water moves slowly or on the outside of a river curve where the water moves faster?

12. Why is most of the sediment deposited in the Delta rather than farther upstream?

13. When sediment hardens or consolidates, what kinds of rock are formed?

14. What is erosion?

15. What things can cause erosion?

16. How are erosion and sedimentation different?

17. How can erosion be slowed or stopped?

18. Why is it important to take care of topsoil in an area?

19. Suppose a rock breaks into several small pieces after a big rock falls on it. Is this an example of physical weathering or chemical weathering? Explain your answer.

20. Give some examples of how rocks can undergo physical weathering.

Short Answer Questions: (5 points each)

1. When farmers dig wells to get water to drink, how deep do the wells have to be to reach water?

2. The amount of spaces between rocks, sand, or other earth materials compared to the total volume of the material is known as what?

3. What is a water aquifer?

4. Are most oil and gas deposits found near the surface of the ground or deep below the ground?

5. When water has filled all the pores in a porous underground area, the area is said to be _____ with water.

6. What determines how deep water soaks into the ground?

7. Give an example of how underground rocks can be weathered physically.

8. Briefly explain how stalactites and stalagmites form in caves.

9. Is the water that erupts from a geyser in Yellowstone National Park hot or cold?

10. Where does the water come from that is spewed out of a geyser?

11. When is a sinkhole most likely to fall in — when there has been plenty of rain or during a drought?

12. How do glaciers "pluck" rocks as they move?

13. Are glaciers able to transport rocks and other materials from place to place?

14. Are glacial deposits primarily pushed ahead of moving glaciers or moved by the process of plucking?

15. Are till deposits and moraines laid down by glaciers, wind, or water?

16. Briefly describe the two kinds of glaciers.

17. Why do rocks attached to glaciers start out jagged and end up being worn down and rounded?

18. Humus makes up a part of the top layer of soil. How is humus made?

19. Suppose a farmer loses most of the topsoil in a field as a result of using poor conservation methods. Will the soil be replaced by natural processes if he doesn't plant another crop on it the next year?

20. Can adding a bag of fertilizer to a field (where topsoil has been lost) make it as good as it was before the topsoil was lost?

Short Answer Questions: (2 points each)

1. Which lines go from the North Pole to the South Pole?

2. Which lines circle the earth and are parallel to the equator?

3. What is a GPS device? What can a GPS device in an automobile do?

4. The materials in the lower part of the mantle are extremely hot, but they are thought to be in a plastic state in places. Under what conditions does this plastic behave like a solid?

5. Under what conditions does material from the mantle rise up into the crust or even to the surface of the earth?

6. The three main states of matter are solid, liquid, and gas. What is meant by a plastic state?

7. Some geologists believe there are convection-type currents in parts of the earth's mantle and core. What are some ways these currents might affect the earth?

8. What are some of the natural resources found in the earth's crust?

9. What is happening where the Pacific Plate and the North American Plate meet along the coast of California?

10. Where is the Mid-Atlantic Ridge?

11. What do geologists believe about a land mass known as Pangaea?

12. According to the elastic rebound theory, what often causes the ground to shake during an earthquake?

13. The strength of an earthquake is reported as a number from 0 to 10. What is this scale of numbers called?

14. What is the name of the instrument that is used to study and identify earthquakes?

15. Which earthquake is more powerful — one that measures 3 on the Richter scale or one that measures 9 on the Richter scale?

16. When magma reaches the surface of the earth, what is it called?

17. What has probably happened if a volcano suddenly erupts and forcefully throws out volcanic materials and rocks?

18. Where are some of the most likely places to find volcanoes in the world?

19. Which of the following processes is most likely to produce flat, level layers of sediment — glaciers, wind, water, or volcanic eruptions?

20. When magma reaches the surface of the earth and hardens there, what kind of rock forms?

21. What are intrusive rocks?

22. How does the cooling rate affect the size of mineral crystals that form in intrusive magma?

23. What kind of map shows contour lines, as well as other details of specific features in an area?

24. Is the increase or decrease in height between any two contour lines on a map the same?

25. How can you tell the difference in a crystal and a rock?

26. Why is granite classified as a rock instead of a mineral?

27. Are minerals and crystals pure substances or mixtures of chemicals?

28. There are three main groups of rocks — sedimentary, metamorphic, and igneous. In which group would limestone be classified? In which group would marble be classified?

29. What conditions are generally necessary for metamorphic rocks to form?

30. Explain how to do a streak test on a rock.

31. Which group of rocks contains fossils?

32. Which kind of rocks are formed when igneous rocks or sedimentary rocks are subjected to high temperatures and/or pressure?

33. Name three or four sedimentary rocks.

34. Basalt, granite, and obsidian are examples of what kind of rock?

35. What is the process of depositing sediment out of water?

36. Is the settling rate faster for large/heavy materials or for small/lighter materials?

37. When sediment hardens or consolidates, what kinds of rock are formed?

38. What is erosion?

39. What things can cause erosion?

40. When farmers dig wells to get water to drink, how deep do the wells have to be to reach water?

41. The amount of spaces between rocks, sand, or other earth materials compared to the total volume of the material is known as what?

42. What is a water aquifer?

43. When water has filled all the pores in a porous underground area, the area is said to be _____ with water.

44. When is a sinkhole most likely to fall in — when there has been plenty of rain or during a drought?

45. Why do rocks attached to glaciers start out jagged and end up being worn down and rounded?

46. Suppose a farmer loses most of the topsoil in a field as a result of using poor conservation methods. Will the soil be replaced by natural processes if he doesn't plant another crop on it the next year?

47. Where are earthquakes most prone to occur?

48. How do glaciers "pluck" rocks as they move?

49. What determines how deep water soaks into the ground?

50. Humus makes up a part of the top layer of soil. How is humus made?

Quiz and Test Answers

for Use with

Science Starters: Elementary Physical & Earth Science

Forces & Motion ⚊● Quiz Answer Keys Level 1 & 2

Quiz 1 Level 1, chapters 1-4

1. time
2. no
3. centimeters – second
4. frame of reference
5. north
6. no
7. friction
8. friction
9. heat
10. reduce
11. yes
12. streamlined
13. yes
14. yes
15. sliding friction
16. engineer
17.

Possible reconstruction #1 Possible reconstruction #2

Quiz 2 Level 1, chapters 5-8

1. inertia
2. yes
3. sea level
4. yes
5. yes
6. air
7. gravity
8. yes
9. yes
10. yes
11. volume
12. buoyant

13. no
14. density
15. lift
16. yes
17. same, air, resistance, falls

Quiz 3 Level 1, chapters 9-14

1. inertia
2. no
3. no
4. no
5. force
6. gravity
7. friction
8. no
9. torsion
10. Newton
11. no
12. yes
13. gravity
14. no
15. space
16. reaction
17. matter, moving, speed, friction

Quiz 4 Level 1, chapters 15-20

1. lower
2. yes
3. no
4. faster
5. moveable
6. yes
7. yes
8. big
9. wheel
10. no

Quiz 1 Level 2, chapters 1-4

1. c
2. a
3. c
4. c
5. a
6. c
7. b

8. a
9. a
10. c
11. c
12. b
13. b
14. b
15. b
16. c
17. b
18. b
19. a
20. c

Quiz 2 Level 2, chapters 5-8

1. c
2. a
3. c
4. a
5. a
6. a
7. a
8. b
9. b
10. a
11. c
12. c
13. b
14. c
15. a
16. b
17. b
18. b
19. a
20. a

Quiz 3 Level 2, chapters 9-14

1. c
2. c
3. b
4. b
5. a
6. a
7. b
8. b

9. b
10. a
11. c
12. a
13. c
14. b
15. a
16. b
17. c
18. c
19. b
20. a
21. a
22. a
23. b
24. c
25. a

Quiz 4 Level 2, chapters 15-20

1. a
2. b
3. a
4. a
5. a
6. a
7. c
8. a
9. b
10. a
11. b
12. b
13. a
14. b
15. b
16. a
17. b
18. a
19. a
20. a
21. c
22. a
23. b
24. b
25. b

Forces & Motion ━●
Test Answer Key
Level 1 & 2

Test 1 Level 1

1. time
2. frame of reference
3. friction
4. friction
5. heat
6. reduce
7. yes
8. yes
9. inertia
10. yes
11. yes
12. air
13. gravity
14. engineer
15. buoyant
16. density
17. lift
18. Bernoulli
19. inertia
20. no
21. force
22. gravity
23. no
24. gravity
25. space
26. yes
27. no
28. faster
29. moveable
30. no
31. Answers will vary. Should mention at least four of the following points for full credit:
- There are upward and downward forces acting on the pencil as it falls.
- The pencil's weight (from the earth's gravity) is a force that pulls down on the pencil.

- There is some friction between the pencil and the air, which pushes up on the pencil.
- The upward force on the pencil is increased by connecting a parachute-like device to the pencil. The parachute produces more air resistance.
- More than one force was acting on the pencil as it was falling.
- The weight of the pencil (gravitational force) was pulling down while the air resistance/friction was pushing up.
32. Answers will vary. Should mention at least four of the following points for full credit:
- The thing that determines whether or not the ship will float is the weight of the water it displaces.
- If the weight of the displaced water is equal to the weight of the ship, the ship will float.
- If the weight of the displaced water is less than the weight of the ship, the ship will sink.
- The upward force the water exerts on floating or sunken boats is called buoyant force.
- Recall that when two forces oppose each other, they are said to be balanced when they cancel each other out. There are balanced forces acting on a floating ship.
- The weight of the ship (from the earth's gravitational force) pulls down.
- At the same time, the buoyant force on the ship pushes up. Therefore, the two forces cancel each other out, and the forces acting on the boat are balanced.

Test 1 Level 2

1. c
2. c
3. c
4. a
5. a

6. c
7. c
8. b
9. a
10. a
11. a
12. c
13. a
14. a
15. a
16. a
17. b
18. a
19. c
20. b
21. b
22. c
23. c
24. b
25. a
26. b
27. b
28. a
29. c
30. c
31. a
32. c
33. a
34. a
35. a
36. b
37. b
38. c
39. b
40. a
41. a
42. a
43. a
44. c
45. a
46. b
47. a
48. a
49. b
50. b

Bonus question: c

The Earth 🔑
Quiz Answer Keys
Level 1 & 2

Quiz One, Level 1 Chapters 1-4

1. longitudinal line
2. the Greenwich Meridian or the Prime Meridian
3. longitudinal lines
4. latitudinal lines
5. the part of the earth on or near the equator
6. 24
7. it stands for Global Positioning System.
8. crust
9. iron and nickel
10. under land
11. convection-type
12. Any of the following: In addition to rocks and soil, the crust also contains water, coal, oil, gas, ores, and mineral deposits.
13. seafloor spreading
14. earthquakes
15. Himalayan Mountains
16. Pangaea
17. elastic rebound theory
18. less
19. San Andreas Fault
20. Ring of Fire
21. Global Positioning Systems can tell you where you are on the earth. They are usually able to show you how to get to a specific address.
22. Earthquakes tend to occur along the boundary of crustal plates or along other fault lines, especially where one plate in the earth's crust is pushing against another plate or where two plates are sliding past each other.
23. The two plates are sliding and grinding past each other as the Pacific Plate moves northward at the rate of about 5 centimeters (just over an inch and a half) per year.
24. Something in-between a solid and a liquid; often depends on the amount of pressure and temperature on the material.
25. The engineering design and the materials used in buildings

Quiz Two, Level 1 Chapters 5-8

1. Richter scale
2. seismograph
3. every day
4. one that measures 9 on the Richter scale
5. lava
6. yes
7. volcanoes
8. no
9. Grand Canyon, Badlands of South Dakota, other places
10. Any of these: water, wind, glaciers, and volcanic eruptions
11. water
12. faults
13. extrusive rocks
14. intrusive
15. intrusive
16. They could enforce building codes for houses, schools, and other buildings that reduce the chances of the buildings falling in during an earthquake. Even if everyone could not afford to do this, the government could be sure that certain buildings, such as schools and hospitals, were built according to safe building codes.
17. One belt follows the coastline around the Pacific Ocean and is known as the "Ring of Fire." The other belt is next to the Mediterranean Sea and extends to southern Asia.

18. Surtsey

Quiz Three, Level 1 Chapters 9-12

1. no
2. steep
3. gradual
4. topographic
5. no
6. small
7. yes
8. minerals
9. quartz, mica, and feldspar.
10. silicate
11. feldspars, micas, olivines, pyroxenes, amphiboles, quartz, clay minerals, and calcite (calcium carbonate)
12. calcium carbonate
13. limestone
14. sedimentary
15. heat and pressure
16. plants and animals that live in the ocean (or once lived there)
17. the surface of the ocean (sea level)
18. Granite is made up of several minerals that have been cemented together into a rock. A mineral is a pure substance.
19. Crystals have flat faces and a definite shape and rocks don't. Crystals are pure substances and rocks are a mixture of minerals (crystals).
20. color test, streak test, luster, crystal form, cleavage test, hardness test (Mohs's scale of hardness), density test, test for magnetic properties, fluorescent glow under ultraviolet light, radioactivity, formation of bubbles when exposed to weak acid, flame test, and others

Quiz Four, Level 1 Chapters 13-16

1. sedimentary, igneous, and metamorphic
2. metamorphic
3. intrusive igneous
4. sandstone, limestone, shale, conglomerate, and others
5. igneous
6. sedimentation
7. deposition
8. large/heavy
9. Delta

10. glaciers
11. on hills/bare ground
12. oxbow lake
13. physical weathering
14. expands
15. sharp and jagged
16. topsoil
17. sedimentation
18. iron oxide
19. limestone
20. Erosion is a broad term that includes the processes that move soil, sediment, and other materials on the earth from one place to another.

Quiz Five, Level 1 Chapters 17-20

1. water table
2. dry
3. pores
4. porosity
5. aquifer
6. deep below
7. saturated
8. runoff
9. stalactites; stalagmites
10. hot
11. drought
12. glaciers
13. valley/continental
14. humus
15. no
16. The porosity of the soil. If there is a layer of soil that is not porous, such as clay, then water cannot get past that layer (unless there are cracks in the clay layer). As long as the soil is porous (or cracked), it can seep farther down.
17. Even though water can't seep past a layer of hard rock, if the rock is cracked, water can travel through the cracks and cause a wearing down of the rocks.
18. When water combines with carbon dioxide, a weak acid known as carbonic acid forms. Carbonic acid reacts chemically with limestone rocks, causing the limestone to come apart.
19. In some places around the world, streams and shallow wells are dangerously polluted with disease-causing organisms, but are the only available source of drinking water. Water from clean wells would greatly improve the health of people in these areas.

Quiz 1, Level 2 Chapters 1-4

1. longitudinal line
2. the Greenwich Meridian or the Prime Meridian
3. longitudinal lines
4. latitudinal lines
5. the part of the earth on or near the equator
6. It stands for Global Positioning System. It can tell you where you are on the earth. It is usually able to show you how to get to a specific address.
7. when they are under great pressure and very high temperatures
8. when there are cracks in the earth's crust
9. Something in-between a solid and a liquid. It often depends on the amount of pressure and temperature on the material.
10. They might cause parts of the earth's crust to slowly move. They may play a role in producing the earth's magnetic field.
11. In addition to rocks and soil, the crust also contains water, coal, oil, gas, ores, and mineral deposits.
12. The two plates are sliding and grinding past each other as the Pacific Plate moves northward at the rate of about 5 centimeters (just over an inch and a half) per year.
13. It runs north and south between North America and Europe and between South America and Africa.
14. seafloor spreading.
15. collided into each other, but neither plate slid under the other one.
16. Pangaea may have been the original land mass that broke apart and formed today's continents.
17. The rocks are bent and are under tension until they break loose. There is a brief period when the rocks vibrate and shake until they reach a state that is free from tension.
18. They tend to occur along the boundary of crustal plates or along other fault lines, especially where one plate in the earth's crust is pushing against another plate or where two plates are sliding past each other.
19. The Pacific Plate is slowly moving past the North American Plate.
20. No. These faults are not at the boundary of two plates, but are deep cracks in the crust beneath the surface.

Quiz 2, Level 2 Chapters 5-8

1. the Richter scale
2. seismograph
3. every day
4. one that measures 9 on the Richter scale
5. Structures designed in this way do not absorb as much of the energy from the earthquake as structures that are tightly attached to the earth's foundation.
6. New Madrid earthquake in 1811; San Francisco earthquake in 1906; Anchorage, Alaska earthquake in 1964. (There are others also.)
7. One belt follows the coastline around the Pacific Ocean and is known as the "Ring of Fire." The other belt is next to the Mediterranean Sea and extends to southern Asia.
8. lava
9. Steam and other gases probably built up inside the mountain until the pressure became great enough to push through a containing wall.
10. Volcanoes, like earthquakes, tend to be found where large crustal plates meet. Sometimes they are found in areas known as "hot spots."
11. No, sometimes magma hardens underneath the surface of the earth.
12. Grand Canyon, Badlands of South Dakota, other places
13. water
14. They may be changed by processes such as folding, tilting, or uplifting.
15. faults
16. through cracks in the crust
17. extrusive rocks
18. rocks that cool and harden from hot magma below the surface of the earth
19. the stratified layers of rocks
20. Crystals are formed when the magma cools slowly. The crystals are much smaller when the magma cools rapidly.

Quiz 3, Level 2 chapters 9-12

1. It means that every place along that line is 990 feet above sea level.
2. no
3. topographic map
4. yes
5. the surface of the ocean (sea level)
6. no
7. Crystals have flat faces and a definite shape and rocks don't. Crystals are pure substances and rocks are a mixture of minerals (crystals).

8. yes

9. minerals

10. yes (same shape, but not necessarily the same size)

11. minerals

12. Granite is made up of several minerals that have been cemented together into a rock. A mineral is a pure substance.

13. Minerals and crystals are pure substances.

14. color test, streak test, luster, crystal form, cleavage test, hardness test (Mohs's scale of hardness), density test, test for magnetic properties, fluorescent glow under ultraviolet light, radioactivity, formation of bubbles when exposed to weak acid, flame test, and others

15. silicate minerals (composed of silicon and oxygen chemically joined together)

16. (1) Limestone is most often a whitish or gray color. (2) You can rub a copper penny forcefully across it and it will leave a scratch mark. (3) You can put a limestone rock in a container of vinegar and it will make bubbles.

17. calcium carbonate

18. Limestone is classified as a sedimentary rock. Marble is classified as a metamorphic rock.

19. heat and pressure

20. It can be eroded by acids. (Carbonic acid and vinegar are common acids in homes.)

Quiz 4, Level 2 Chapters 13-16

1. Rub a rock on a piece of unglazed ceramic tile and observe if it leaves a colored streak. Try to match the color of the streak with what is shown in a reference book or Internet source.

2. sedimentary, igneous, and metamorphic

3. sedimentary

4. metamorphic

5. sandstone, limestone, shale, conglomerate, and others

6. igneous

7. limestone

8. sedimentation (or deposition)

9. wind and glaciers

10. large/heavy

11. inside of a curve

12. The water is moving very slowly as it enters the Gulf. Moving currents of water are able to carry sediment long distances.

13. sedimentary rocks

14. Erosion is a broad term that includes the processes that move soil, sediment, and other materials on the earth from one place to another.

15. Erosion can be caused by water, mudflows, ice, wind, and gravity.

16. Erosion is a process that moves soil and other materials away and sedimentation is a process that leaves deposits of sediment behind.

17. by covering ground with vegetation, especially on hilly areas

18. It may take hundreds of years to produce a good, thick layer of topsoil, and it can be lost in a few years by erosion.

19. This is physical weathering, because the pieces of rock all have many of the same properties they had before they broke.

20. Exposure to moving water, wind, ice; grinding against other rocks; changing temperatures cause rocks to expand when they are heated and contract when they are cooled; freezing of water in cracks in rocks; growth of plants; landslides.

Quiz 5, Level 2 Chapters 17-20

1. Wells must reach the water table, the place where water has saturated a layer of rocks, sand, and other soil material.

2. porosity

3. a rock formation with enough porosity to hold water and allow water to be moved in and out of the rock

4. deep below the ground

5. saturated

6. The porosity of the soil. If there is a layer of soil that is not porous, such as clay, then water cannot get past that layer (unless there are cracks in the clay layer). As long as the soil is porous (or cracked), it can seep farther down.

7. Even though water can't seep past a layer of hard rock, if the rock is cracked, water can travel through the cracks and cause a wearing down of the rocks.

8. As the water drops from the ceiling, small amounts of calcium carbonate are deposited. These deposits form slowly from the ceiling as stalactites or rise from the floor as stalagmites.

9. hot

10. It moves up through cracks in the rocks from deep below the surface near a source of magma.

11. during a drought

12. Melted glacier water seeps down into cracks in the rocks below the glacier. As the water refreezes, it

expands the cracks, eventually breaking off pieces of rocks. Some of these broken pieces of rock may freeze to the bottom of the glacier.

13. yes
14. moved by the process of plucking
15. by glaciers
16. Valley glaciers form in and move down valleys to seek sea level. Continental glaciers are in the form of huge sheets of ice that move over plains and low, hilly land.
17. The newly broken rocks that are plucked up by a moving glacier tend to be jagged, but as they grind and rub against the land, they get worn down.
18. It is made of decayed, weathered tree leaves, limbs, or other once-living things.
19. No. The formation of soil is a process that takes many years.
20. No. It would help, but a bag of fertilizer can't provide the same benefits as topsoil.

The Earth ⚷
Test Answer Key

Test 1 Level 1
1. The Greenwich Meridian or the Prime Meridian
2. longitudinal
3. latitudinal
4. iron/nickel
5. Pangaea
6. elastic rebound
7. Ring of Fire
8. Richter scale
9. seismograph
10. lava
11. extrusive
12. intrusive
13. no
14. steep
15. topographic
16. yes
17. minerals
18. feldspars, micas, olivines, pyroxenes, amphiboles, quartz, clay minerals, and calcite
19. heat and pressure
20. sedimentary, igneous, and metamorphic
21. sandstone, limestone, shale, conglomerate, and others
22. sediment

23. deposition
24. physical weathering
25. water table
26. porosity
27. aquifer
28. runoff
29. continental/valley
30. humus
31. Crystals have flat faces and a definite shape and rocks don't. Crystals are pure substances and rocks are a mixture of minerals (crystals).
32. In some places around the world, streams and shallow wells are dangerously polluted with disease-causing organisms, but are the only available source of drinking water. Water from clean wells would greatly improve the health of people in these areas.

Test 1 Level 2
1. longitudinal lines
2. latitudinal lines
3. It stands for Global Positioning System. It can tell you where you are on the earth. It is usually able to show you how to get to a specific address.
4. when they are under great pressure and very high temperatures
5. when there are cracks in the earth's crust
6. Something in-between a solid and a liquid. It often depends on the amount of pressure and temperature on the material.
7. They might cause parts of the earth's crust to slowly move. They may play a role in producing the earth's magnetic field.
8. In addition to rocks and soil, the crust also contains water, coal, oil, gas, ores, and mineral deposits.
9. The two plates are sliding and grinding past each other as the Pacific Plate moves northward at the rate of about 5 centimeters (just over an inch and a half) per year.
10. It runs north and south between North America and Europe and between South America and Africa.
11. Pangaea may have been the original land mass that broke apart and formed today's continents.
12. The rocks are bent and are under tension until they break loose. There is a brief period when the rocks vibrate and shake until they reach a state that is free from tension.
13. the Richter scale
14. seismograph

15. one that measures 9 on the Richter scale

16. lava

17. Steam and other gases probably built up inside the mountain until the pressure became great enough to push through a containing wall.

18. Volcanoes, like earthquakes, tend to be found where large crustal plates meet. Sometimes they are found in areas known as "hot spots."

19. water

20. extrusive rocks

21. rocks that cool and harden from hot magma below the surface of the earth

22. Larger crystals are formed when the magma cools slowly. The crystals are much smaller when the magma cools rapidly.

23. topographic map

24. yes

25. Crystals have flat faces and a definite shape and rocks don't. Crystals are pure substances and rocks are a mixture of minerals (crystals).

26. Granite is made up of several minerals that have been cemented together into a rock. A mineral is a pure substance.

27. Minerals and crystals are pure substances.

28. Limestone is classified as a sedimentary rock. Marble is classified as a metamorphic rock.

29. Heat and pressure

30. Rub a rock on a piece of unglazed ceramic tile and observe if it leaves a colored streak. Try to match the color of the streak with what is shown in a reference book or Internet source.

31. sedimentary

32. metamorphic

33. sandstone, limestone, shale, conglomerate, and others

34. igneous

35. sedimentation (or deposition)

36. large/heavy

37. sedimentary rocks

38. Erosion is a broad term that includes the processes that move soil, sediment, and other materials on the earth from one place to another.

39. Erosion can be caused by water, mudflows, ice, wind, and gravity.

40. Wells must reach the water table, the place where water has saturated a layer of rocks, sand, and other soil material.

41. porosity

42. a rock formation with enough porosity to hold water and allow water to be moved in and out of the rock

43. saturated

44. during a drought

45. The newly broken rocks that are plucked up by a moving glacier tend to be jagged, but as they grind and rub against the land, they get worn down.

46. No. The formation of soil is a process that takes many years.

47. Earthquakes tend to occur along the boundary of crustal plates or along other fault lines, especially where one plate in the earth's crust is pushing against another plate or where two plates are sliding past each other.

48. Melted glacier water seeps down into cracks in the rocks below the glacier. As the water refreezes, it expands the cracks, eventually breaking off pieces of rocks. Some of these broken pieces of rock may freeze to the bottom of the glacier.

49. The porosity of the soil. If there is a layer of soil that is not porous, such as clay, then water cannot get past that layer (unless there are cracks in the clay layer). As long as the soil is porous (or cracked), it can seep farther down.

50. It is made of decayed, weathered tree leaves, limbs, or other once-living things.

Master Supply List

for Use with

Science Starters: Elementary Physical & Earth Science

Forces & Motion
Semester Supply List
Common Household Items

- ❏ 10-cm diameter circular disc cut from cardboard or poster board
- ❏ 3 x 5 index card
- ❏ Ball
- ❏ Balloons
- ❏ Black marker
- ❏ Board for ramp
- ❏ Books for weight
- ❏ Broom handle or other vertical bar
- ❏ Chart
- ❏ Clock or watch with second hand
- ❏ Doll
- ❏ Flathead metal screws (same size)
- ❏ Flathead screwdriver (big and small handle)
- ❏ Flour (five-pound bag)
- ❏ Funnel
- ❏ Glue
- ❏ Hardback book (2-4 pounds)
- ❏ Heavy string or cord
- ❏ Heavy-duty rubber band
- ❏ Lid for plastic water bottle
- ❏ Liquid detergent
- ❏ Long sheet of paper
- ❏ Measuring stick/tape measure (metric with each cm divided into tenths)
- ❏ Metal washers or nuts (4 to 6)
- ❏ Missile (small, stale marshmallows work well)
- ❏ Newspaper
- ❏ Overflow cup (or a container with a spout)
- ❏ Paper
- ❏ Paper clip (6 small, 1 large)
- ❏ Paper towels
- ❏ Pencils, (1 with an eraser, 1 round)
- ❏ Pennies (16 minted after 1982)
- ❏ Piece of scrap carpet
- ❏ Plastic cup and tub or bucket with handle
- ❏ Popsicle stick (or something similar)
- ❏ Quart jar
- ❏ Ramp
- ❏ Safety glasses
- ❏ Scissors
- ❏ Straw
- ❏ String or fishing line
- ❏ Table
- ❏ Tape (regular and double sided)
- ❏ Timer or watch with second hand
- ❏ Toothpicks
- ❏ Toy car that rolls easily
- ❏ Yardstick or thin wooden strip (unfinished)

Additional Household Items

- ❏ Balance scales if available
- ❏ C clamp (to clamp board to table)
- ❏ Feathers
- ❏ Film canister with lids or similar
- ❏ Graduated cylinder
- ❏ Hook screws with a small circular end, at least 12 cm in length (2)
- ❏ Insulated wire, single-strand, 30 cm long
- ❏ Metal hook (optional)
- ❏ Ping-pong ball
- ❏ Roller blades or skates
- ❏ Rubber stopper, one-hole (or similar)
- ❏ Sandpaper (long enough to wrap around the wooden block)
- ❏ Sewing spool (2 large empty)
- ❏ Skateboard
- ❏ Spring force measure or fish scales
- ❏ Sturdy plastic tube (like the cylinder of an old plastic ink pen or a piece of plumbing tubing)
- ❏ Styrofoam balls or other balls (2)
- ❏ Toy wagon
- ❏ Various flathead metal screws of same length
- ❏ Various wooden boards/blocks
- ❏ Walking toy

List courtesy of: **InvestigateThePossibilities.org**

Visit the site for more information and specialty items.

Forces & Motion Supply List by Investigation

Investigation #1: Wind-up Walking Toys
Gather These Things:

- ❏ Walking toy
- ❏ Chart
- ❏ Pencil
- ❏ Clock with second hand
- ❏ Toothpicks
- ❏ Metric ruler (with each cm divided into tenths)

Investigation #2: Which Way Did It Go?
Gather These Things:

- ❏ Toy wagon
- ❏ Walking toy
- ❏ Long, smooth board
- ❏ Broom handle or other vertical bar

Investigation #3: Investigating Friction?
Gather These Things:

- ❏ Heavy string
- ❏ Spring scale (force measure)
- ❏ Round pencils
- ❏ Hardback book (2-4 pounds)

Investigation #4: Friction — Does It Rub You the Wrong Way?
Gather These Things:

- ❏ Smooth board (about 60 cm long)
- ❏ Stack of books
- ❏ Toy car that rolls easily
- ❏ Metric ruler or tape measure
- ❏ Piece of scrap carpet
- ❏ Wooden block
- ❏ Piece of scrap wood
- ❏ A piece of sandpaper (long enough to wrap around the wooden block)

Investigation #5: That's Heavy, Dude — Air Pressure
Gather These Things:

- ❏ Small plastic cup (4 to 8 ounce)
- ❏ Large index card
- ❏ Sheet of paper
- ❏ Plastic tub or bucket

For Optional Activity

- ❏ Sheet of newspaper
- ❏ Table
- ❏ Safety glasses
- ❏ A thin yardstick or thin wooden strip (unfinished)

Investigation #6: Floating Pencil Race — Gravity Versus Air Resistance
Gather These Things:

- ❏ Pencil
- ❏ Paper
- ❏ Tape
- ❏ Scissors
- ❏ Paper towels
- ❏ Feathers
- ❏ String or line
- ❏ Watch (with a second hand)

Investigation #7: What Floats Your Boat?
Gather These Things:

- ❏ Overflow cup (or a container with a spout)
- ❏ Empty film canister with lids
- ❏ 16 pennies (minted after 1982)
- ❏ Graduated cylinder
- ❏ Quart jar
- ❏ Water
- ❏ Liquid detergent
- ❏ Balance scales if available

Investigation #8: Giving Airplanes a Lift
Gather These Things:

- ❏ String
- ❏ Sheet of paper
- ❏ Funnel
- ❏ Two styrofoam balls (or other balls)
- ❏ Ping-pong ball
- ❏ Glue

Investigation #9: Crash Test Dummies — Inertia and That Sudden Stop
Gather These Things:

- ❏ Skateboard
- ❏ Ball
- ❏ 3 x 5 card
- ❏ Penny
- ❏ Doll
- ❏ Tape

- ❏ Long sheet of paper
- ❏ Plastic cup
- ❏ Small wooden blocks
- ❏ Toy car that rolls
- ❏ Board for ramp
- ❏ Small ruler

Investigation #10: Cars and Ramps — What Does Newton Have to Say?
Gather These Things:

- ❏ Small car
- ❏ Ramp
- ❏ Four blocks
- ❏ Measuring tape
- ❏ Chart

Investigation #11: The Mighty Conquering Catapults
Gather These Things:

- ❏ Two hook screws with a small circular end, at least 12 cm in length
- ❏ Heavy-duty rubber band (about 1 cm thick)
- ❏ Pencil or marker
- ❏ Lid for plastic water bottle (no more than 1 cm deep)
- ❏ Double-sided mounting tape (for hanging mirrors and pictures)
- ❏ A 5cm by 10cm by 30 cm (2" by 4" by 12") piece of soft pine board
- ❏ A lightweight object to serve as a missile (small, stale marshmallows work well)
- ❏ C clamp (to clamp board to table)
- ❏ One popsicle stick
- ❏ Tape measure

Investigation #12: Round and Round without Stopping
Gather These Things:

- ❏ Sturdy plastic tube (like the cylinder of an old plastic ink pen or a piece of plumbing tubing)
- ❏ Small weights
- ❏ Heavy-duty string
- ❏ One-hole rubber stopper (or something similar)
- ❏ Metal hook (optional)

Investigation #13: Roller Derby with Flour — Action and Reaction
Gather These Things:

- ❏ Student on roller blades or skates
- ❏ Five-pound bag of flour
- ❏ Metric ruler or tape measure
- ❏ Chalk

Investigation #14: Balloon Jet Propulsion — Action and Reaction
Gather These Things:

- ❏ Pencil
- ❏ Balloons
- ❏ String or fishing line
- ❏ Straw
- ❏ Tape

Investigation #15: Balancing Act with a Stick — Center of Gravity and Mass
Gather These Things:

- ❏ Popsicle stick (or something similar)
- ❏ Four to six metal washers or nuts
- ❏ Ruler
- ❏ Single-strand insulated wire, 30 cm long
- ❏ Sharpened pencil

Investigation #16: Spinning Tops — Rotational Inertia and Mass
Gather These Things:

- ❏ Pencil or a small dowel rod
- ❏ 10-cm diameter circular disc cut from cardboard or poster board (with every 60 degrees marked)
- ❏ Timer or watch with second hand
- ❏ Six paper clips (same size)
- ❏ Six pennies
- ❏ Tape
- ❏ Scissors

Investigation #17: He Ain't Heavy, He's Just My Load, Brother
Gather These Things:

- ❏ Board one meter long with positions marked
- ❏ Triangular block
- ❏ Books for weight
- ❏ Spring force measure or fish scales
- ❏ Heavy string or cord

Investigation #18: How Do Like Your Pulleys —
Fixed, Moving, or Combined?
Gather These Things:

- ❏ Two large empty thread spools
- ❏ Small bucket with handle containing a weight (load)
- ❏ Spring force measure
- ❏ Heavy cord
- ❏ Measuring stick or tape measure
- ❏ 30 cm solid core insulated wire

Investigation #19: And the Wheel Goes Round
Gather These Things:

- ❏ A large empty sewing spool
- ❏ Pencil with an eraser
- ❏ A weight
- ❏ Large paper clip
- ❏ Heavy string
- ❏ Small-handle flathead screwdriver
- ❏ Big-handle flathead screwdriver
- ❏ Flathead metal screws (same size)
- ❏ Board with two hook screws (used for Activity #11)
- ❏ C-clamp if needed to hold board to table
- ❏ Scrap wood for screws

Investigation #20: If It Doesn't Move, How Can It
Be a Machine?
Gather These Things:

- ❏ Flat board about 40 cm long for a ramp
- ❏ Spring force measure
- ❏ Object to move (about five pounds)
- ❏ Stack of books
- ❏ Paper
- ❏ Pencil
- ❏ Scissors
- ❏ Black marker
- ❏ Flathead screwdriver
- ❏ Scrap wood pieces
- ❏ Various kinds of flathead metal screws of same length

The Earth
Semester Supply List
Common Household Items

- ❏ Aluminum foil
- ❏ Books
- ❏ Clear glasses
- ❏ Colored markers
- ❏ Copper penny
- ❏ Cornstarch
- ❏ Cup
- ❏ Disposable aluminum pie pan
- ❏ Empty egg carton (with 12 holders)
- ❏ Eyedropper
- ❏ Flat map of the world
- ❏ Fruit peeler
- ❏ Glue
- ❏ Hardboiled egg
- ❏ Hard pieces of candy
- ❏ Heavy paper (or cardboard)
- ❏ Knives
- ❏ Loaf bread (or layers of cake)
- ❏ Magnifying glass
- ❏ Map of the world
- ❏ Masking tape
- ❏ Measuring cup
- ❏ Metal washers (3)
- ❏ Nail to punch holes in the pan
- ❏ Notebook
- ❏ Orange
- ❏ Paper
- ❏ Paper cup
- ❏ Paper plates
- ❏ Paper towels (for cleanup, etc.)
- ❏ Peanut butter and jelly (or dark icing)
- ❏ Pencil and pen
- ❏ Plastic quart-size jar with screw-on lid
- ❏ Popsicle stick
- ❏ Safety glasses (for all)
- ❏ Salt
- ❏ Scissors
- ❏ Soft drink can, aluminum
- ❏ Sticks
- ❏ Small plastic bags that can be closed easily
- ❏ Stirrer (wooden or old spoon)
- ❏ String
- ❏ Sturdy rubber band

- ❏ Tape
- ❏ Tape measure
- ❏ Thread
- ❏ Tin cans, same size, with lids removed (6)
- ❏ Toothpicks, colored
- ❏ Various containers and jars
- ❏ Vinegar
- ❏ Wire strainer

Additional Household Items

- ❏ Alum, 3 ounces
- ❏ Ceramic tile (unglazed side)
- ❏ Effervescent tablets
- ❏ Film canisters or similar items
- ❏ Garden spade for digging
- ❏ Globe (if available)
- ❏ Iron nail
- ❏ Limestone rocks (garden/landscaping item)
- ❏ Modeling clay, 4 colors
- ❏ Reference books/Internet on the structure of the earth, locating places, major tectonic plates
- ❏ Rock and mineral identification chart
- ❏ Rolled wafers with a dark filling (or narrow flat wafers)
- ❏ Sandpaper
- ❏ Small, dry twigs, leaves, pine straw, or mulch
- ❏ Soil, potting, and sand
- ❏ Steel wool pad
- ❏ Styrofoam
- ❏ Topographic map showing contour layers
- ❏ Variety of rock and minerals
- ❏ Various blocks of wood

List courtesy of: **InvestigateThePossibilities.org**
Visit the site for more information and specialty items.

The Earth Supply List by Investigation

Investigation #1: Orange You Going to Map the Earth?
Gather These Things:
- ❏ Whole orange
- ❏ Fruit peeler
- ❏ Plastic knife
- ❏ Black marker
- ❏ Green marker
- ❏ Flat map of the world
- ❏ Globe (if available)

Investigation #2: Composition of the Earth
Gather These Things:
- ❏ Modeling clay in four colors
- ❏ Toothpicks
- ❏ Cornstarch
- ❏ Water
- ❏ Glue
- ❏ Strips of paper
- ❏ Shallow pan or bowl
- ❏ Measuring cup
- ❏ Paper towels (for cleanup)
- ❏ Reference book on the structure of the earth

Investigation #3: Why Is Everything Moving?
Gather These Things:
- ❏ Map of the world
- ❏ Internet or other resources about plate tectonics
- ❏ Scissors
- ❏ Colored marker
- ❏ Metal knife (not sharp)
- ❏ Hardboiled egg

Investigation #4: Earthquake
Gather These Things:
- ❏ Large piece of coarse sandpaper
- ❏ Wooden block covered with coarse sandpaper with a hook screwed into one end (about 10 cm in length — longer is okay)
- ❏ Sturdy rubber band
- ❏ Masking tape
- ❏ Clear adhesive tape
- ❏ Popsicle stick
- ❏ Heavy string or cord
- ❏ Tape measure

- ❏ Safety glasses (for all observers)
- ❏ Heavy paper (or cardboard)
- ❏ Fine sandpaper (optional)

Investigation #5: Living with Earthquakes
Gather These Things:
- ❏ Map of the world
- ❏ Red marker
- ❏ Yellow marker
- ❏ References for finding locations of places on the earth
- ❏ References for finding the major tectonic plates on the earth
- ❏ Safety glasses (for the optional Part B)

Investigation #6: Volcanoes
Gather These Things:
- ❏ Several film canisters
- ❏ Effervescent tablets
- ❏ Water
- ❏ Cup to pour from

Investigation #7: Mountains (Folding and Faulting)
Gather These Things:
- ❏ Four colors of clay
- ❏ Four layers of Styrofoam
- ❏ Colored markers
- ❏ Pencil

Investigation #8: Pardon the Intrusion
Gather These Things:
- ❏ Three or four slices of loaf bread with crust trimmed off (or layers of cake)
- ❏ Peanut butter and jelly mixed together (or dark icing)
- ❏ Butter knife
- ❏ Knife to cut the iced layers
- ❏ Colored toothpicks
- ❏ Rolled wafers with a dark filling (or narrow flat wafers)

Investigation #9: Mapping a Mountain
Gather These Things:
- ❏ Clay (enough to make a model of a mountain; see recipe in teacher's book)
- ❏ Small plastic ruler with metric units
- ❏ A few blocks of wood, all the same thickness
- ❏ Paper

- ❑ A copy of a topographic map showing contour lines
- ❑ Marker

Investigation #10: Growing Crystals
Gather These Things:

- ❑ Three ounces of powdered alum
- ❑ Three pint-size glass containers
- ❑ Three small clear glasses
- ❑ Hot water
- ❑ Stirrer
- ❑ Three pieces of thread, about 30 cm
- ❑ Three small metal washers (nuts or rocks will also work)
- ❑ Tape
- ❑ Magnifying lens
- ❑ Dark paper

Investigation #11: Minerals
Gather These Things:

- ❑ Variety of rocks and minerals
- ❑ Small plastic bags that can be closed easily
- ❑ Permanent marker
- ❑ Magnifying lens

Investigation #12: Rocks That Fizz
Gather These Things:

- ❑ Limestone rocks (You may be able to find limestone rocks because they are common rocks in many places. They can also be purchased in plant nurseries or garden shops.)
- ❑ Various other kinds of small rocks
- ❑ Eyedropper
- ❑ Paper plates
- ❑ Small clear containers
- ❑ Vinegar
- ❑ Water
- ❑ Paper towels
- ❑ Magnifying lens

Investigation #13: Rocks Have an ID
Gather These Things:

- ❑ Backside of ceramic tile (unglazed side)
- ❑ Collection of rocks (at least 12)
- ❑ White paper
- ❑ Colored paper
- ❑ Copper penny
- ❑ Iron nail

- ❑ Empty egg carton (with 12 holders)
- ❑ Writing pen
- ❑ Rock and mineral identification charts on the Internet or in other references to help identify or classify the rocks in your collection.

Investigation #14: How Little, Tiny Things Settle Out of Water to Become Rocks
Gather These Things:

- ❑ Quart jar with lid
- ❑ Soil
- ❑ Gravel
- ❑ Sand
- ❑ Very small, dry pieces of twigs
- ❑ Water

Investigation #15: How Rocks and Dirt Catch a Ride
Gather These Things:

- ❑ Rectangular pan or cookie sheet
- ❑ Aluminum foil (optional)
- ❑ Paper cup and a toothpick to poke holes in it
- ❑ Water
- ❑ Sand or dirt
- ❑ Books
- ❑ Paper towels
- ❑ Large pan
- ❑ Small, dry twigs, leaves, pine straw, or mulch

Investigation #16: Physical and Chemical Weathering
Gather These Things:

- ❑ Empty aluminum soft drink can
- ❑ Disposable, plastic quart-size jar with screw-on lid
- ❑ Water
- ❑ Small margarine tubs
- ❑ Salt
- ❑ Vinegar
- ❑ Handful of rocks from outside
- ❑ Several pieces of hard candy (rounded or cylindrical, at least 1 cm thick with no outer coating)
- ❑ Limestone rocks (garden or landscaping store)
- ❑ Steel wool pad (remove any soap)

Investigation #17: Holes in Rocks
Gather These Things:

- ❑ Six clean empty tin cans, same size, with lids removed
- ❑ Container about three or four times larger than the paper cup
- ❑ Dry sand
- ❑ Wire strainer
- ❑ Small to medium size rocks
- ❑ Metric liquid measuring cup
- ❑ Small paper cup
- ❑ Sturdy toothpick

Investigation #18: Caves, Sinkholes, and Geysers
Gather These Things:

- ❑ Disposable aluminum pie pan
- ❑ Clay made from flour, salt, water, and oil (see teacher's book)
- ❑ Nail to punch holes in the pan
- ❑ Potting soil
- ❑ Plastic knife
- ❑ Container to catch water

Investigation #19: Glaciers
Gather These Things:

- ❑ Small plastic container
- ❑ Medium plastic container
- ❑ Large flat container
- ❑ Water
- ❑ Sand and dirt
- ❑ Small rocks
- ❑ Paper towels

Investigation #20: Toiling in the Soil
Gather These Things:

- ❑ Garden spade for digging
- ❑ Magnifying glass
- ❑ Four sturdy sticks
- ❑ String
- ❑ Pencil
- ❑ Metric tape measure
- ❑ Notebook

List courtesy of: **InvestigateThePossibilities.org**
Visit the site for more information and specialty items.